University-Industry Technology Transfer in the UK:

Emerging Research and Opportunities

Martin Wynn
University of Gloucestershire, UK

A volume in the Advances in Knowledge Acquisition, Transfer, and Management (AKATM) Book Series

IGI Global
DISSEMINATOR OF KNOWLEDGE

Published in the United States of America by
IGI Global
Information Science Reference (an imprint of IGI Global)
701 E. Chocolate Avenue
Hershey PA, USA 17033
Tel: 717-533-8845
Fax: 717-533-8661
E-mail: cust@igi-global.com
Web site: http://www.igi-global.com

Library of Congress Cataloging-in-Publication Data

Names: Wynn, Martin, author.
Title: University-industry technology transfer in the UK : emerging research and opportunities / by Martin Wynn.
Description: Hershey, PA : Information Science Reference, an imprint of IGI Global, [2019] | Includes bibliographical references and index.
Identifiers: LCCN 2018027793| ISBN 9781522574088 (hardcover) | ISBN 9781522574095 (ebook)
Subjects: LCSH: Technology transfer--Great Britain. | Universities and colleges--Great Britain--Graduate work. | Business and education--Great Britain. | Higher education and state--Great Britain. | University of Gloucestershire.
Classification: LCC T174.3 .W97 2019 | DDC 338.941/06--dc23 LC record available at https://lccn.loc.gov/2018027793

This book is published in the IGI Global book series Advances in Knowledge Acquisition, Transfer, and Management (AKATM) (ISSN: 2326-7607; eISSN: 2326-7615)

British Cataloguing in Publication Data
A Cataloguing in Publication record for this book is available from the British Library.

Advances in Knowledge Acquisition, Transfer, and Management (AKATM) Book Series

ISSN:2326-7607
EISSN:2326-7615

Editor-in-Chief: Murray E. Jennex, San Diego State University, USA

MISSION

Organizations and businesses continue to utilize knowledge management practices in order to streamline processes and procedures. The emergence of web technologies has provided new methods of information usage and knowledge sharing.

The **Advances in Knowledge Acquisition, Transfer, and Management (AKATM) Book Series** brings together research on emerging technologies and their effect on information systems as well as the knowledge society. **AKATM** will provide researchers, students, practitioners, and industry leaders with research highlights surrounding the knowledge management discipline, including technology support issues and knowledge representation.

COVERAGE

- Cognitive Theories
- Cultural Impacts
- Information and Communication Systems
- Knowledge acquisition and transfer processes
- Knowledge management strategy
- Knowledge Sharing
- Organizational Learning
- Organizational Memory
- Small and Medium Enterprises
- Virtual communities

IGI Global is currently accepting manuscripts for publication within this series. To submit a proposal for a volume in this series, please contact our Acquisition Editors at Acquisitions@igi-global.com or visit: http://www.igi-global.com/publish/.

The Advances in Knowledge Acquisition, Transfer, and Management (AKATM) Book Series (ISSN 2326-7607) is published by IGI Global, 701 E. Chocolate Avenue, Hershey, PA 17033-1240, USA, www.igi-global.com. This series is composed of titles available for purchase individually; each title is edited to be contextually exclusive from any other title within the series. For pricing and ordering information please visit http://www.igi-global.com/book-series/advances-knowledge-acquisition-transfer-management/37159. Postmaster: Send all address changes to above address.

Titles in this Series

For a list of additional titles in this series, please visit:
https://www.igi-global.com/book-series/advances-knowledge-acquisition-transfer-management/37159

The Role of Knowledge Transfer in Open Innovation
Helena Almeida (University of Algarve, Portugal) and Bernardete Sequeira (University of Algarve, Portugal)
Information Science Reference • ©2019 • 397pp • H/C (ISBN: 9781522558491) • US $195.00

Scholarly Content and Its Evolution by Scientometric Indicators Emerging Research...
Zahid Ashraf Wani (University of Kashmir, India) and Tazeem Zainab (University of Kashmir, India)
Information Science Reference • ©2019 • 201pp • H/C (ISBN: 9781522559450) • US $135.00

Technological Innovations in Knowledge Management and Decision Support
Nilanjan Dey (Techno India College of Technology, India)
Information Science Reference • ©2019 • 322pp • H/C (ISBN: 9781522561644) • US $195.00

Effective Knowledge Management Systems in Modern Society
Murray E. Jennex (San Diego State University, USA)
Information Science Reference • ©2019 • 391pp • H/C (ISBN: 9781522554271) • US $195.00

Global Information Diffusion and Management in Contemporary Society
Zuopeng (Justin) Zhang (State University of New York at Plattsburgh, USA)
Information Science Reference • ©2019 • 331pp • H/C (ISBN: 9781522553939) • US $195.00

Contemporary Knowledge and Systems Science
W. B. Lee (Hong Kong Polytechnic University, China) and Farzad Sabetzadeh (Hong Kong Polytechnic University, China)
Information Science Reference • ©2018 • 340pp • H/C (ISBN: 9781522556558) • US $185.00

Enhancing Knowledge Discovery and Innovation in the Digital Era
Miltiadis D. Lytras (The American College of Greece, Greece) Linda Daniela (University of Latvia, Latvia) and Anna Visvizi (Effat University, Saudi Arabia)
Information Science Reference • ©2018 • 363pp • H/C (ISBN: 9781522541912) • US $215.00

For an entire list of titles in this series, please visit:
https://www.igi-global.com/book-series/advances-knowledge-acquisition-transfer-management/37159

IGI Global
DISSEMINATOR OF KNOWLEDGE

701 East Chocolate Avenue, Hershey, PA 17033, USA
Tel: 717-533-8845 x100 • Fax: 717-533-8661
E-Mail: cust@igi-global.com • www.igi-global.com

Table of Contents

Foreword

All but ignored in the time of Adam Smith, the central role of technology and innovation to successful business growth is widely acknowledged today. Building on its expertise in basic science, the UK as a nation has a superb track record in developing and applying technologies in industries such as pharmaceuticals, advanced manufacturing, aerospace and automobiles.

In these and other areas the role of the UK university sector has been especially marked. The UK rightly has a global reputation in higher education, and over recent decades universities have proven increasingly adept at working with small and medium sized (SME) companies, often as part of complex integrated supply chains.

But elsewhere, the picture is more mixed. UK manufacturing productivity lags behind its principle European competitors, and much of this can be traced to a long tail of less innovative SMEs. SMEs as a whole play a key role in the UK economy, as employers and as producers, but as many studies have shown the sector has a long way to go to meet the best international standards in management, human capital development and adoption of new technology.

Given all this, the publication of this book by Martin Wynn, a well-known academic/business expert practitioner, is especially welcome. It focuses in detail on the use of Knowledge Transfer Partnerships in companies across nineteen detailed case studies, reviewing the experience of those involved in each case, defining outcomes and analysing strengths and weaknesses, both at the individual level and overall.

By adopting a regional focus based on Gloucestershire and Herefordshire, the author has been able to assess a cross-section of real-world cases that act both as a snapshot of one specific area and as a wider reference point for SMEs in general.

The book covers a wide range of mainstream technologies from customer relationship management, enterprise resource planning and e-procurement systems, to bespoke developments for e-business operations and new product development. It can be read with great interest by business professionals and practitioners involved in technology transfer projects.

Jesse Norman
MP, UK

Jesse Norman *has been the Member of Parliament for Hereford and South Herefordshire since 2010. Educated at Oxford University (BA) and University College London (MPhil, PhD) he was a Director at Barclays before entering politics, and combines wide business and academic experience.*

Preface

Since the turn of the century, technology transfer and innovation have played an increasingly important role in UK Government policy for re-invigorating and supporting British industry, and technology transfer from third parties such as universities is often seen as a catalyst for achieving increased competitiveness, particularly for small to medium sized enterprises (SMEs). SMEs are defined in a European context as having less than 250 staff (European Commission, 2016), including small business enterprises (SBEs) which have between 10 and 49 staff. A lack of financial resources and basic technological capability can act as barriers to SMEs adopting new technologies, both for their in-house systems or in the incorporation of new technologies into their products or services (Guzzini & Iacobucci, 2017).

Brychan (1999) underlined the importance of technology transfer networks for SMEs, particularly those where technology is transferred into an SME from an external source, and the term "open innovation" was first used by Chesbrough (2003) to denote the use of external resources as part of the research and development process for new technology. This gave impetus to the harnessing of external capabilities to achieve swifter and more effective results in the application of new technologies in industry. However, related research has often focused on larger companies, and "small and medium-sized enterprises are excluded from the mainstream discussion on open innovation" (Brunswicker & Vanhaverbeke, 2015, p.1241). This is particularly relevant in the UK where there are 5.4 million SMEs, providing over 60% of all private sector employment, with a combined turnover of £1.8 trillion. According to Close Brothers (2011) "over half of SMEs in the UK have invested in new technology and software, and one in four plan to do so in the future" (p.11). In addition, they found that SMEs are "most positive about technology's ability to improve staff efficiency (50%) and help upskill staff (43%)" (p.11).

For all sizes of organisation, the external environment for technology is rapidly evolving. This necessitates regular upgrades to information systems and technical infrastructure and adjustments to future information technology strategies, and technology transfer involving support from third parties is one way of achieving this. However, as Bozeman (2000) concludes, this is not a straightforward process. He notes that:

"anyone studying technology transfer understands just how complicated it can be. First, putting a boundary on 'the technology' is not so easy. Second, outlining the technology transfer process is virtually impossible because there are so many concurrent processes. Third, measuring the impacts of transferred technology challenges scholars and evaluators, requiring them to reach deep down into their research technique kit bag" (p.627).

This book examines the introduction of new technologies into companies partnering with the University of Gloucestershire within the framework of the UK's Knowledge Transfer Partnership (KTP) scheme (discussed below in the Introduction section). The technology in question is not transferred from the university, nor was it developed within the university. Rather, the skills and expertise from university staff are used to lead and manage the introduction of new technologies in the partner companies. It thus falls within Roessner's (2000) conceptualisation of technology transfer as the movement of know-how, technical knowledge, or technology from one organizational setting to another. These projects involve the transfer of know-how and technical knowledge, which is then used to introduce new technologies, either co-developed with the partner company or acquired from third party technology vendors. The term "technology transfer project" is thus used with the qualification that the transfer is more about knowledge, know-how and expertise relating to technology, rather than the technology itself.

UNIVERSITIES AND KNOWLEDGE TRANSFER

The concept of knowledge transfer is not new (Decter, 2009), but there is now a major interest in its role in promoting economic growth and job creation. Hardhill and Baines (2009) concluded that "the promotion of knowledge transfer to maximise public investment has been a recurrent theme in UK policy documents" (p.82) and the university sector has played a significant role in knowledge transfer to industry in the UK for several decades. Lee et al. (2010) found that universities were a valuable source of knowledge and innovation for start-up companies, but also for established businesses where knowledge transfer can play a crucial role in their organisational development. In addition, knowledge transfer and innovation often have spill-over effects that can create wider benefits to the overall economy. The KTP scheme is of national significance in the UK in providing a financial and management framework for effecting such change, with government funding being made available to support companies that wish to take advantage of the wide range of expertise located within universities.

In a European context, the KTP scheme is a useful model, as it is generally agreed that similar mechanisms have not, as yet, been as well embedded in university-industry linkages as in the UK. The European Union's Commissioner of Enterprise and Industry has noted that "European universities and other public research organisations need to engage more actively in the exploitation of publicly-funded research results. It is necessary in order to stimulate innovation and maximise the benefits of publicly funded research, so we can turn scientific research into new products and services, which will create new industries and jobs" (EurActiv, 2008, p.2). There are, nevertheless, voluntary guidelines in place for European universities and industry to engage in knowledge transfer (Commission of the European Communities, 2007), but it is in the UK in particular that collaborative relationships between universities and businesses have received growing attention from governments and academics.

A number of government commissioned reports and reviews have explored ways of extending and enhancing these relationships and have made a wide range of recommendations designed to further that end. The Lambert Review of Business-University Collaboration (Lambert, 2003) acknowledged the scale of public investment in teaching and research within the UK's universities and formally endorsed the belief that "transferring the knowledge and skills between universities and business and the wider community increases the economic and social returns" (p.31). The Sainsbury Review (Department for Innovation, Universities and Skills, 2007) concluded that "while our knowledge partnerships seem to be working well for our research universities, there is scope to increase knowledge transfer from business-facing universities to small and medium sized enterprises" (p.60). It was in the same year that Chesbrough discussed and defined the open innovation concept that the Department of Trade and Industry specified a range of products for promoting and enabling knowledge transfer and innovation, in particular to support technology transfer to SMEs (Department of Trade and Industry, 2003). Now, in 2018, the UK Government is preparing to establish a "knowledge exchange framework" for universities and collaborating companies. One of the supporting research documents in this initiative (Higher Education Funding Council for England,

2016) reports that "evidence points to the UK university system operating at world class standard in technology transfer practice" (p.4).

The organisational and broader economic and social benefits of academic engagement with the business world have also been explored within academic literature in the UK and in many other parts of the world (Ankrah et al., 2013; de Wit-de Vries et al., 2018; Perkmann et al., 2013; Puerta Sierra & Jasso Villazul, 2018). A variety of collaborative relationships have been studied, but nevertheless, Wang and Lu (2007) concluded that there was "a lack of in-depth study on the process of knowledge transfer between university and industry, in particular how knowledge is transferred across institutional boundaries between university and industry during the interactive process" (p.120). It is this interface that this book addresses, particularly in the context of SMEs, who "often miss several important activities associated with the technology transfer that are critical for the quick deployment and success of the new technologies being transferred" (Jagoda, Maheshwari, & Lonseth, 2010, p. 366).

TECHNOLOGY TRANSFER AND INNOVATION

Technology transfer will normally involve innovation to some degree and this has been a recurrent theme in government support for small firms. Twenty years ago, Hoffman, et al. (1998) noted "of particular interest is the way in which SMEs innovate, and much recent policy-making has been directed at mechanisms to support this activity" (p.40). They found that there were many individual countries operating SME support schemes – for example, the Business Links programme in the UK which provided access to innovation and technology counsellors to help SMEs with innovation related problems. These support networks were the forerunners of today's KTP scheme and more recently, the Sainsbury Review (Department for Innovation, Universities and Skills, 2007) identified the need "to make the most of publicly funded research and to increase innovation in business and public services" (p.60). In 2013, the Technology Strategy Board, one of the main funding bodies for the KTP scheme, stressed the importance of innovation in securing KTP funding (Technology Strategy Board, 2013); and in a different national context, Silva, Gaia, Caten and Facó (2017) found that "cooperation between industry-university can increase in a significant way the capability of enterprises' innovation" (p.49).

However, there are different definitions and varying degrees of innovation. There is a common distinction in existing literature between radical innovation and incremental innovation. The former normally involves the introduction of fundamental changes, often in the technology sphere, that are linked to a company's long-term business objectives, and often take many years to fully materialise and deliver expected benefits. Indeed, Pedersen and Dalum (2004) suggest that this often represents a new technological paradigm. Incremental innovation, on the other hand, though it may well involve the application of new technology to deliver significant organisational benefits, is normally delivered within a 6-24 month period, and this aligns with the technology change introduced in the projects discussed in this book. Urabe (1988) defines innovation as "the generation of a new idea and its implementation into a new product, process or service" (p.3). Similarly, Popadiuk and Choo (2006) suggest that "innovation consists of new ideas that have been transformed or implemented as products, processes or services, generating value for the firm" (p.308).

Other authors (Lichtenthaler, 2011; Robertson, Casali, & Jacobson, 2012) have concluded that external knowledge sourcing requires certain internal capabilities for the effective integration and application of new knowledge. Brunswicker and Vanhaverbeke (2015) suggest that "so far, little is known about the role of such integrative managerial practices for innovation in external knowledge sourcing in SMEs" (p.1242), and Enkel, Bell and Hogenkamp (2011) have noted that "we still lack a clear understanding of these mechanisms..... and how we can gain maximum advantage from this approach" (p.1162). Alvarez and Iske (2015), in their study of 142 Dutch SMEs, empirically analyse possible complementarity or substitutability between internal capabilities and external knowledge sourcing. Their findings "suggest a negative interplay between internal capabilities and external knowledge sourcing" (p.55). Chesbrough (2003) differentiated between two concepts of open innovation: inbound, where new ideas flow into an organization, and outbound, where internally developed technologies and ideas can be acquired by external organizations. As noted by Brunswicker and Vanhaverbeke (2015), "to successfully benefit from inbound open innovation, a firm requires some higher-order management capabilities to align inbound knowledge flows with the firm's in-house innovation activities" (p.1243). Guzzini and Iacobucci (2017) analysed the factors affecting the likelihood of the failure of innovation projects, and the relation between project failure and innovation performance. Based on data from German firms in the period 2002–2005, they highlighted the significance of collaboration with universities and public research institutions as a key factor in determining successful outcomes.

KNOWLEDGE MANAGEMENT

There is a clear link in the literature between knowledge management, innovation and technology transfer. For example, Gloet and Samson (2016) researched the relationship between knowledge management and innovation and concluded that "the management of knowledge may indeed hold the key to increasing systematic innovation capability in organizational contexts", and that "for managers, this involves developing new forms of knowledge, embedding this new knowledge within organizations, as well as managing flows of information, knowledge and experience" (p.55). Nevertheless, Marouf and Khalil (2015) point out that "the knowledge management (KM) literature in general is short on field evidence concerning knowledge sharing (KS) practices in project management settings, where knowledge occupies a central place" (p.1).

This is reinforced by Wu, Hsu and Yeh (2007) who note that "although the importance of knowledge transfer has been recognized for years, the determinants of knowledge transfer still remain unclear" (p.335). In their study of sales teams in the e-travel industry, they identified two important determinants of knowledge transfer - knowledge sharing and learning intensity. They concluded that "overall, the results support the argument that social capital facilitates the determinants of knowledge transfer" and that "this finding is robust at the team level" (p.335). Similarly, Silva et al. (2013), in their study of technology transfer and knowledge management in a Brazilian technological innovation centre, found that "innovation and knowledge management processes in a company can be understood as a cluster of commitments generated by individuals aiming to create, acquire, transform, apply and subsequently protect the creation of knowledge generated by the organization in order to remain competitive in an active market" (p.80). They concluded, however, that "there is no single model for creating and managing organizational knowledge in all organizations that seek technology transfer, as each organization has its characteristics and a different culture from one another" (p.86).

Rafiei, Akhavan and Hayati (2016) researched the role of knowledge management in successful technology transfer in the Iranian aerospace industry in an attempt to identify the key factors and variables affecting the assimilation of technology. Their findings indicated "that there is a significant and positive relationship between knowledge management and successful technology transfer effectiveness" and that "relational and organizational capabilities - as key factors and facilitators - play a mediating role between knowledge management and technology transfer" (p.178). In the same vein, Nahar, Al-Obaidi and Huda (2001), in looking at knowledge management in international technology transfer, found that "in most cases, the transfer of technology requires the transfer of knowledge related to physical process contained in physical elements (e.g. computer) as well as the knowhow related to operating them" (p.356).

PURPOSE AND ORGANISATION OF THE BOOK

This book examines how technology transfer has operated in university-company projects undertaken in SMEs via the KTP scheme. It adopts a qualitative case study approach, focusing on 19 different cases drawn from companies in the regions adjoining the University of Gloucestershire, UK. A model of twelve change factors that underpin successful outcomes is used to assess and compare these projects. The projects examine how new technologies were introduced to promote innovation in internal processes, in services to customers, or in new product development. This builds on existing theory and models in the field of study, and highlights the significance of these change factors for the effective delivery of technology transfer projects in SMEs.

The related concepts of technology transfer, innovation and knowledge management have been discussed in recent literature (Mention & Asikainen, 2012; Striukova & Rayna, 2015), but much of the empirical evidence available to date has focussed on high technology industries, often in large multinational companies (Vanhaverbeke, Vermeersch, & De Zutter, 2012). The model discussed in this book makes a contribution to theory and practice in the field of technology transfer for SMEs that are not necessarily advanced technology users. Brunswicker and Vanhaverbeke (2015) identified the dearth of knowledge regarding "internal capabilities for managing innovation" and concluded that "little is known about the role of such integrative managerial practices for innovation in external knowledge sourcing in SMEs" (p.1242).

Moshonsky, Serenko and Bontis (2014) noted that "academic knowledge is only relevant to industry if it motivates practitioners to take action inspired by its content" and that "future research examining the transfer of academic knowledge to practice should focus on knowledge transfer mechanisms" (p.71). The model put forward in this book provides an analytical framework that will be of interest and value to academics and business practitioners looking to develop university-industry partnerships involving technology change and innovation. It can be used as a framework for periodic monitoring and review of technology transfer projects. An assessment of progress and competence relating to the twelve change factors provides a comprehensive overview of project status which can help maintain the desired balance between the different dimensions of multi-faceted projects.

A recent report by the Confederation of British Industry (CBI) concluded that British businesses are falling behind their European competitors by failing to adopt established technologies, a trend which is undermining productivity and harming growth and living standards. The report noted that there is "low-hanging fruit" worth as much as £100 billion of gross value per annum, and that companies needed to invest in well-established systems to catch up with more advanced countries. The report (Confederation of British Industry, 2017) found that "failure to adopt tried-and-tested technologies and management best practices" was a "a major driver" (p.10) of low productivity, and Carolyn Fairbairn (as cited in Wallace, 2017, para.5), the CBI's director-general, concluded that "the UK is a decade behind Denmark - the proportion of companies with e-purchasing, enterprise resource planning [systems] is below the level where Denmark was in 2009."

This book examines how technology transfer projects such as those cited by the CBI have been accomplished in a range of industry sectors. It does not attempt to make generalisations about these sectors, but rather provides examples from these sectors. Nor is this book about the commercialisation of technology developed in universities by third parties. It is about how companies, and particularly SMEs, can successfully implement proven technologies to support their growth and development.

The book is organized into seven chapters. The first six chapters look at case examples of technology transfer projects in different industry sectors, and each chapter uses the change factors assessment model discussed below in the Introduction section.

Chapter 1 concerns the *contract packaging industry* and examines three projects that had varying degrees of success in promoting company growth.

Chapter 2 focuses on two companies in the *construction sector*, both house builders, but operating in different sub-sectors. Although the business models underpinning these companies' operations were very different, both succeeded in implementing a range of new systems to support process change and bottom-line growth.

Chapter 3 examines four projects in the *software development industry*, three of which concerned new product development. The cases illustrate the problems faced by small software companies when introducing new technology products into a dynamic and competitive marketplace.

Chapter 4 looks at four projects in the *service industries sector*, comprising environmental, training, project management and financial transaction services companies. New customer relationship management systems were introduced in three of the four companies in an attempt to improve customer facing processes.

Chapter 5 reviews three projects in two companies from the *product assembly, sales and marketing* sector. One company is in the office furniture and design business, where improved management information from its legacy systems was viewed as critical for company growth. The second company, who sell a range of products for the elderly and disabled, adopted a new information systems strategy and then implemented bespoke e-procurement systems. The projects exhibit varying degrees of success.

Chapter 6 looks at three manufacturing companies. Two of these companies looked to implement integrated enterprise resource planning systems to support process change and improved management information, whilst the third company adopted a "best of breed" strategy that linked shop floor data capture systems with point solutions for transaction processing and management reporting.

Chapter 7 concludes with a review and analysis of all nineteen projects included in the book and discusses the relevance of this research for current and the future technology challenges.

Martin Wynn
University of Gloucestershire, UK

REFERENCES

Alvarez, H., & Iske, P. (2015). Internal Capabilities and External Knowledge Sourcing for Product Innovation in LMT SMEs. *Journal of Innovation Management*, *3*(2), 55–70.

Ankrah, S. N., Burgess, T. F., Grimshaw, P., & Shaw, N. E. (2013). Asking both the university and industry actors about their engagement in knowledge transfer: What single-group studies of motives omit. *Technovation*, *33*(2/3), 50–65. doi:10.1016/j.technovation.2012.11.001

Bozeman, B. (2000). Technology transfer and public policy: A review of research and theory. *Research Policy*, *29*(4-5), 627–655. doi:10.1016/S0048-7333(99)00093-1

Brunswicker, S., & Vanhaverbeke, W. (2015). Open Innovation in Small and Medium-Sized Enterprises (SMEs): External Knowledge Sourcing Strategies and Internal Organizational Facilitators. *Journal of Small Business Management*, *53*(4), 1241–1263. doi:10.1111/jsbm.12120

Brychan, T. (1999). A model of diffusion of technology into SMEs. In *Naples Conference Proceedings, 1999.* International Council for Small Business. Retrieved February 14th, 2014 from www.sbaer.uca.edu/research/icsb/1999/TOC.pdf

Chesbrough, H. W. (2003). *Open Innovation - the new imperative for creating and profiting from technology*. Boston, MA: Harvard Business School Press.

Close Brothers. (2017). *The Power of Productivity*. London: Close Brothers.

Commission of the European Communities. (2007). *Voluntary guidelines for universities and other research institutions to improve their links with industry across Europe. Accompanying Document to the Communication from the Commission, Improving Knowledge Transfer between Research Institutions and Industry across Europe: Embracing Open Innovation– Implementing the Lisbon Agenda*. Brussels: SEC.

Confederation of British Industry. (2017). *From Ostrich to Magpie - Increasing Business Take-Up of Proven Ideas and Technologies.* Retrieved September 19th, 2018 from http://www.cbi.org.uk/index.cfm/_api/render/file/?method=inline&fileID=F840035B-0678-4A72-BAF8B7D170F47A07

de Wit-de Vries, E., Dolfsma, W. A., van der Windt, H. J., & Gerkema, M. P. (2018). Knowledge transfer in university--industry research partnerships: A review. *The Journal of Technology Transfer*. doi:10.100710961-018-9660-x

Decter, M. H. (2009). Comparative review of UK-USA industry-university relationships. *Education + Training*, *51*(8/9), 624–634. doi:10.1108/00400910911005190

Department for Innovation, Universities, and Skills. (2007). *The Race to the TOP: A Review of government's Science and Innovation Policies*. Retrieved April 15th, 2017 from http://www.rsc.org/images/sainsbury_review051007_tcm18-103118.pdf

Department of Trade and Industry. (2003). *DTI Innovation Report: Competing in the global economy: the innovation challenge*. Retrieved October 18th, 2014 from http://www.saiie.co.za/ocs/public/conferences/1/schedConfs/1/program-en_US.pdf#page=46

Enkel, E., Bell, J., & Hogenkamp, H. (2011). Open Innovation Maturity Framework. *International Journal of Innovation Management*, *15*(6), 1161–1189. doi:10.1142/S1363919611003696

Euractiv. (2008). *EU Pushes for Better Knowledge Transfer*. Retrieved July 15th, 2012 from www.euractiv.com/ en/science/eu-pushes-better-knowledge-transfer/article-171564

European Commission. (2016). *User guide to the SME Definition*. Document Ref. Ares (2016) 956541 - 24/02/2016. Retrieved September 20th, 2018 from http://ec.europa.eu/regional_policy/sources/conferences/state-aid/sme/smedefinitionguide_en.pdf

Gloet, M., & Samson, D. (2016). Knowledge Management and Systematic Innovation Capability. *International Journal of Knowledge Management*, *12*(2), 54–72. doi:10.4018/IJKM.2016040104

Guzzini, E., & Iacobucci, D. (2017). Project failures and innovation performance in university–firm collaborations. *The Journal of Technology Transfer*, *42*(4), 865–883. doi:10.100710961-016-9554-8

Hardhill, I., & Baines, S. (2009). Personal reflections on Knowledge Transfer and changing UK research priorities. *21st Century Society, 4*(1), 81-94

Higher Education Funding Council for England. (2016). *University Knowledge Exchange (KE) Framework: good practice in technology transfer.* Report to the UK higher education sector and HEFCE by the McMillan group. Higher Education Funding Council for England. Retrieved September 20th, 2018 from http://dera.ioe.ac.uk/27123/1/2016_ketech.pdf

Hoffman, K., Parejo, M., Bessant, J., & Perren, L. (1998). Small firms, R&D, technology and innovation in the UK: A literature review. *Technovation*, *18*(1), 39–55. doi:10.1016/S0166-4972(97)00102-8

Jagoda, K., Maheshwari, B., & Lonseth, R. (2010). Key issues in managing technology transfer projects: Experiences from a Canadian SME. *Management Decision, 48*(3), 366-382. doi:10.1108/00251741011037747

Lambert, R. (2003). *Lambert Review of Business-University Collaboration – Final Report*. Retrieved May 9th, 2014 from www.lambertreview.org.uk

Lee, N., Morris, K., Wright, J., Jones, A., & Brinkley, I. (2010). *No city left behind? The geography of the recovery and the implications for the coalition*. London: The Work Foundation.

Lichtenthaler, U. (2011). Open Innovation: Past Research, Current Debates, and Future Directions. *The Academy of Management Perspectives*, *25*(1), 75–93.

Marouf, L., & Khalil, O. (2015). The Influence of Individual Characteristics on Knowledge Sharing Practices, Enablers, and Barriers in a Project Management Context. *International Journal of Knowledge Management*, *11*(1), 1–27. doi:10.4018/IJKM.2015010101

Mention, A.-L., & Asikainen, A.-L. (2012). Innovation and productivity: Investigating the effects of openness in services. *International Journal of Innovation Management*, *16*(3), 1–27. doi:10.1142/S136391961240004X

Moshonsky, M., Serenko, A., & Bontis, N. (2014). Examining the Transfer of Academic Knowledge to Business Practitioners: Doctoral Program Graduates as Intermediaries. *International Journal of Knowledge Management*, *10*(3), 70–95. doi:10.4018/ijkm.2014070105

Nahar, N., Al-Obaidi, Z., & Huda, N. (2001). Knowledge Management in International Technology Transfer. In D. Kocaoglu, T. Anderson, D. Milosovic, T. Daim, T. Gulledge, C. Kim, & H. Tschirky (Eds.), *Technology Management in the Knowledge Era* (pp. 355-364). IEEE and PICMET.

Pedersen, C., & Dalum, B. (2004). Incremental versus radical change - the case of the digital north Denmark Program. *International Schumpeter Society Conference*.

Perkmann, M., Tartari, V., McKelvey, M., Autio, E., Brostrom, A., D'Este, P., ... Sobrero, M. (2013). Academic engagement and commercialisation: A review of the literature on university-industry relations. *Research Policy*, *42*(2), 423–442. doi:10.1016/j.respol.2012.09.007

Popadiuk, S., & Choo, C. W. (2006). Innovation and Knowledge Creation – How are these concepts related? *International Journal of Information Management*, *26*(4), 302–312. doi:10.1016/j.ijinfomgt.2006.03.011

Rafiei, A., Akhavan, P., & Hayati, S. (2016). Knowledge management in successful technology transfer (Case study: Iranian aerospace industries and knowledge-based centers). *Aircraft Engineering and Aerospace Technology: An International Journal*, *88*(1), 178–188. doi:10.1108/AEAT-11-2013-0220

Robertson, P., Casali, G., & Jacobson, D. (2012). Managing open incremental process innovation: Absorptive capacity and distributed learning. *Research Policy*, *41*(5), 822–832. doi:10.1016/j.respol.2012.02.008

Roessner, J. D. (2000). Technology transfer. In C. Hill (Ed.), *Science and Technology Policy in the US, A Time of Change*. London: Longman.

Silva, L., Gaia, S., Caten, C., & Facó, R. (2017). Technology Transfer and Innovation Management: The Brazilian TTOs Challenges. *International Journal of Knowledge Management*, *13*(2), 49–64. doi:10.4018/IJKM.2017040104

Silva, L., Kovaleski, J., Gaia, S., Garcia, M., & de Andrade Júnior, P. (2013). Technology Transfer and Knowledge Management in Technological Innovation Center: A Case Study in Brazil. *Journal of Management and Strategy*, *4*(2), 78-87.

Striukova, L., & Rayna, T. (2015). University-industry knowledge exchange: An exploratory study of Open Innovation in UK universities. *European Journal of Innovation Management*, *18*(4), 471–492. doi:10.1108/EJIM-10-2013-0098

Technology Strategy Board. (2013). *Concept to Commercialisation: A strategy for business innovation, 2011-2015*. Technology Strategy Board. Retrieved September 24th, 2018 from https://assets.publishing.service.gov.uk/government/uploads/system/uploads/attachment_data/file/360620/Concept_to_Commercialisation_-_A_Strategy_for_Business_Innovation_2011-2015.pdf

Urabe, K. (1988). Innovation and the Japanese Management System. In K. Urabe, J. Child, & T. Kagono (Eds.), *Innovation and management international comparisons*. Berlin: Walter de Gruyter. doi:10.1515/9783110864519-005

Vanhaverbeke, W., Vermeersch, I., & De Zutter, S. (2012). *Open innovation in SMEs: how can small companies and start-ups benefit from open innovation strategies?* Leuven, Belgium: Flanders District of Creativity.

Wallace, T. (2017). UK firms missing out on £100bn by ignoring tried and tested tech, says CBI. *Daily Telegraph*. Retrieved November 13th, 2017 from http://www.hl.co.uk/news/2017/11/13/uk-firms-missing-out-on-100bn-by-ignoring-tried-and-tested-tech-says-cbi

Wang, Y., & Lu, L. (2007). Knowledge transfer through effective university-industry interactions: Empirical experiences from China. *Journal of Technology Management in China*, *2*(2), 119–133. doi:10.1108/17468770710756068

Wu, W.-L., Hsu, B.-F., & Yeh, R.-S. (2007). Fostering the determinants of knowledge transfer: A team-level analysis. *Journal of Information Science*, *33*(3), 326–339. doi:10.1177/0165551506070733

Acknowledgment

The author acknowledges the role of the Associates, other university personnel, company staff and the UK Government advisors. Without their support and contribution, this book would not have been possible. More specifically, the following colleagues are acknowledged and thanked for their efforts in delivering the technology transfer projects discussed in this book.

Associates

Runu Meah, Oludotun Olubanjo, Claire Holmes, Monika Horner, Toby Henderson, Fazal Elahi, Rauf Butt, Emma Tipton, Hongnian Zang, Matt Barker, Alex Oboubo, Rui Shen, Erin Lau, Phillip Turner, Omar Robb, Rizwan Uppal, Gabriela Maldonado, Hanida Abas, Raul Brandao.

University Staff

Professor Shujun Zhang, Professor Peter Jones, Julie Paterson, Dr David Wakeling, Jon Wise, Andrew Tomlinson, David Liewe, Evelyn Shaw, Caroline Monk

Company Personnel

Peter Roberts, Wayne Kedward, Mark Tooth, Karl Fry, Peter Mayes, Steve Hughes, Tony Sills, Guy Cook, David Brennan, Mike Sadler, Dean Sadler, Paul Bagnall, Paul Morgan, Richard Harris, Robert Oldmeadow, Jane Oldmeadow, Alastair Gibbs, Amanda Harrold, Phillip Turner, Andy Graham, Steve Lesbirel, Paul Westlake, Kerry O'Shea, Simon Oldroyd, Helen Russell, David Abbot, Steve Chittock, Derwyn Roberts, Keith McNeil, Helen Gilbert, Liz Townsend, Simon Kitrick, Mike Clarke, Jonathan Evans, Jeremy Pile,

Ian Hill, Don Howard, Sarah Fenwick, Helen Mistry, Peter Maryszczak, Stuart Hills.

UK Government Advisors

Dr Gillian Rysiecki, Dr David Britton, Dr Ted Jones, Dr Howard Nicholls, Dr Paul Smith

Acknowledgement and thanks are given to the UK Government Department for Business, Energy and Industrial Strategy and its previous incarnations for support for the Knowledge Transfer Partnership (KTP) programmes featured in this book. KTP aims to help businesses to improve their competitiveness and productivity through the better use of knowledge, technology and skills that reside within the UK Knowledge Base. KTP is now funded by Innovate UK along with the other government funding organisations.

Martin Wynn
University of Gloucestershire, UK

Introduction

OVERVIEW

In 2003, the Department of Trade and Industry (DTI) specified a range of products for supporting and promoting innovation, particularly in the field of technology development and application (Department of Trade and Industry, 2003). One of these products was the Knowledge Transfer Partnership (KTP) scheme, which has provided direct support of circa £25m per annum for graduates and their supervisors to undertake specific knowledge transfer projects in firms of all sizes, but particularly in small to medium sized enterprises (SMEs). It was the successor to the Teaching Company Scheme (TCS) that had been in operation since the 1980s (Lipscomb & McEwan, 2001).

The focus of this book is on *technology* transfer in SMEs via the KTP scheme, and this section provides an introduction to some of the key elements explored in the book. As Hannabuss (2001) pointed out, knowledge provides a justification and motivation to alter decisions, and industry practitioners require knowledge to develop and carry out their technology implementation plans. Simmons et al. (2001) established that the process of knowledge transfer mostly fails on the side of the receiver, and Serenko, Bontis and Hull (2011) have argued that future research should focus on knowledge transfer mechanisms and how to improve the overall process.

This book attempts to do precisely that, and in the following sub-section, more information regarding the KTP scheme is provided, along with an overview of the nineteen projects reviewed in the book. This is followed by a review of the research methodology and a discussion of the analytical models used in assessing the projects that feature in the book. Most of these projects were completed some years ago, now allowing a considered assessment of what was of significance in determining their outcome. Indeed, because of the context and impact of some of these projects, it was judged appropriate to wait until now before attempting such a review and analysis.

KNOWLEDGE TRANSFER PARTNERSHIPS

KTP projects are a four-way partnership (Figure 1) between the university, the company, the graduate (or "Associate" as they are termed), and the UK Government sponsoring agency, who provide up to 67% of the required funding. The scheme aims to help businesses to innovate and grow by linking them with a university and a graduate to work on a specific project that will meet a core strategic need with the focus being on delivering increased profits for the business partner. Innovation is a key focus of these projects, which may involve a number of different themes, but the projects studied here concern technology transfer in one form or another. The university recruits and employs a suitably qualified graduate - the Associate - who then works at, and brings new skills and knowledge to, the business. The partnerships involve the Associate working in an organization full time, for a period of between 6 and 36 months, but two years is the normal duration of these projects. During this time, a university academic (the "academic supervisor") is assigned for 25 days per annum to support and supervise the project and to bring in specialist knowledge and expertise as appropriate to ensure project delivery. In addition, the university provides an academic to oversee and generally manage the project from the university side (the "academic lead"). These two roles can be fulfilled by the same person. Training and equipment are also provided, and the total value of the scheme to the SME is circa £70K per annum, of which the SME contributes circa £23K per annum.

There are also many benefits for the partner universities, both to the institution and to the individuals involved in the projects (Wynn & Turner, 2013; Wynn & Jones, 2017); but the focus in this book is mainly on the impact these projects can make in terms of technology in the partner

Figure 1. The four-way partnership that underpins KTP projects

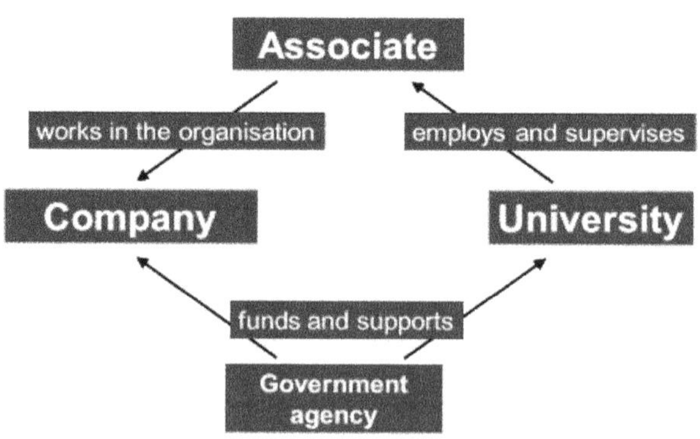

companies. Innovate UK (formerly the Technology Strategy Board, and prior to that, Momenta and the Teaching Company Directorate), now the main UK Government agency responsible for this scheme, has stipulated that it wishes to focus its funding primarily on supporting SMEs in various aspects of innovation, especially those emanating from technology change and competency development (Technology Strategy Board, 2013); and an increasing number of SMEs have embarked upon KTP projects, attracted by the possibilities of growth and innovation in their operations and processes. The funding body for KTP projects also provides a post-project assessment of the project outcomes, providing an A-E grading (A being excellent; E being unsatisfactory).

Since 2003, the University of Gloucestershire has completed 45 KTP projects and in 2013/2014 their Gross Value Added was estimated at £4.1m (Biggar Economics, 2015). Of these 45 projects, 30 were with SMEs, and 30 were related to technology transfer, with the focus being on, for example,

Table 1. Company projects featured in the book

Company Name	Core Activity	Project Duration	Headcount	T/O (£m)
Allpay.net	Financial technology services	2007-9	190	21.8
AuraQ	Process management software & services	2009-11	5	0.5
Beacons Business Interiors	Office design	2004-6	47	6.9
Beaumont Travel	Bus operator/software developer	2005-7	35	0.9
Brecon Pharmaceuticals (1)	Pharmaceutical packaging	2003-5	162	5.6
Brecon Pharmaceuticals (2)	Pharmaceutical packaging	2004-6	231	7.6
Building Solutions	House builder	2005-7	75	5.8
C&G Services	Training services	2005-7	25	1.2
Contrapak	Contract packaging	2009-11	52	1.2
Dowty Propellers	Aircraft blade manufacturer	2004-6	181	N/A
E-Business Services	Web based software developer	2006-8	6	0.2
Energist UK	Environmental consultancy	2009-10	40	1.1
Fixing Point	Roofing materials manufacture	2006-8	53	5.4
Muddy Boots	Supply chain software	2010-12	30	1.3
Optimum Consultancy	Project management services	2008-10	35	2.4
Pegasus Retirement Homes	House builder	2003-5	39	10.4
SKF AeroEngine Bearings	Bearings manufacture	2007-9	250	17
TPG DisableAids (1)	Equipment for elderly & disabled	2005-7	35	2.8
TPG DisableAids (2)	Equipment for elderly & disabled	2009-11	47	4.3

the introduction of new information systems, new e-business technologies, or software product development. This book examines nineteen of these projects in which the author was involved (Table 1). Only at one company – E-Business Services – are aliases used, for reasons of their customer confidentiality agreements. Both the company name and the customer names cited in this case are aliases.

RESEARCH METHODOLOGY

The research underpinning the material presented in this book is based on the well-established qualitative case study approach. Such case studies may be used to develop theory as a result of data analysis (Saunders, Lewis, & Thornhill, 2009). A case study approach allows a "detailed investigation of one or more organizations, or groups within organizations, with a view to providing an analysis of the context and processes involved in the phenomenon under study" (Hartley, 1994, p.323). Case studies "provide the opportunity to place research into a certain context due to the selection of specific sectors, institutions, countries, etc." (Cunningham, Menter, & Young, 2017, p. 923). One of the main strengths of this approach is its depth, and the amount of detail it can generate. Silverman (2013) noted that case studies provide a complex and rich understanding of change projects across a period of time, thus allowing for an appreciation of causality and history, set in a local context. Yin (2012) argued that selection of multiple cases should consider a similar context so that a set of multiple facts can be more easily identified. He suggested that the more the cases, the greater confidence or certainty in a study's findings, and the fewer the cases the less confidence or certainty. The research philosophy contains important assumptions about the way the researcher views the world and therefore influences the research strategy and methods. The research philosophy in the research upon which this book is based can be characterized as interpretivist, based on an inductive approach.

In these case studies, several different methods were used to collect data, all of which are associated with a qualitative approach, including documentation analysis, observation, and interviews. The main documentary sources were as follows: the original project proposals which were written in conjunction with the company managers, and contained project plans and detailed task lists as well as a clear cost-benefit case for the project; the minutes of the weekly project review meetings and of the three monthly Project Board meetings held with the local representative of the UK Government's funding body; the project Final Reports, of which there were two for each project – one jointly authored by the company and the university, and the other written by the Associate (in the following chapters, " Final Report" assumes authoring by the university and company unless otherwise stated as by the Associate); the numerous reports and documents produced by the Associates across the course of these projects (e.g. Quality Plans, Project Briefs); and the many emails sent and received across the duration of these projects and in the time since project closure. These provide a rich source of information that helped identify the key factors in each case study. Participant observation by the academic supervisor has also contributed to the interpretation of events, evidenced both in the formal meetings noted above, but also in the many informal discussions with the Associates and company managers. This provided a deeper understanding of key project issues and why and how decisions were made. These findings were supported by more recent phone conversations with project team members and company management. This allowed further assessment of the projects overall, the decision making involved, and knowledge input processes of team members. Walsham (1995) notes that "it is desirable in interpretive studies to preserve a considerable degree of openness to the field data, and a willingness to modify initial assumptions and theories. This results in an iterative process of data collection and analysis, with initial theories being expanded, revised, or abandoned altogether" (p.76).

Framework analysis (Mason, Mirza, & Webb, 2018) was used to identify the key change factors in these projects. Since it was originally developed by Ritchie and Spencer (1994) for the management of data in applied policy research, this technique has been deployed for qualitative analysis in a number of areas, including research into the psychological (Parkinson et al., 2016), health (Gale et al., 2013) and sociological (Neale, 2012) fields. A series of key steps are followed leading to "highly structured outputs of summarised data" (Gale et al., 2013, p.2). The central output is a matrix table of rows (representing the cases) and columns (of different coded themes or topics). Each cell in the matrix contains summarised data, organised by case and code. This allows for large datasets to be readily viewed and compared, by reading across cases and comparing cases by code.

MODEL DEVELOPMENT

A number of models have been used to assess technology change projects such as those discussed in this book. The stages of growth model (Nolan, 1979), the e-business adoption ladder (Taylor & Murphy, 2003), the Connect-Publish-Interact-Transform model (Department of Trade and Industry, 2004), and the organizational capabilities model (Willcocks & Sauer, 2000; Levy & Powell, 2003) have all been used in various research initiatives to assess such projects. However, these models focus in the main on depicting *what* happened rather than explaining *why* or *how* it happened. Nevertheless, other studies have attempted to identify the key determinants of successful technology transfer. For example, as already noted, Wu, Hsu and Yeh (2007) pointed out the importance of knowledge sharing and learning intensity, and Knockaert et al. (2011) highlight the significance of top management composition in determining project outcomes. The research findings presented in this book attempt to identify the key factors that enabled the successful completion of technology transfer projects within the management and financial framework of the KTP scheme. It builds on some models in the extant literature - for example, the five-stage Innovation Capability Maturity Model (ICMM) of Essman and Du Preez (2009), and its subsequent development by Enkel, Bell and Hogenkamp (2011), who identified the main elements that determined the successful completion of change projects. In their model, the three main elements are Climate for Innovation (covering sub-elements such as clarity of strategy, initiative taking, and clear target assessment), Partnership Capacity (sub-elements of partner satisfaction, network building and diversity in collaboration) and Internal Processes (communication, innovation facilities, knowledge sharing). The analysis of projects in this book builds upon these concepts to develop a new model of twelve change factors that can be used to assess the readiness of SMEs to successfully pursue technology transfer projects, and to ensure projects are kept in balance as implementation proceeds. Three main dimensions of change - relating to people, processes and technology aspects of a project - are used as the overarching framework for assessing these twelve factors. These dimensions of change have been used in the analysis of systems projects (Figure 2), and have also been depicted as the three pillars of success in technology change projects by a number of authors (Department of Trade and Industry, 2004; King-Turner, 2014; Wynn et al., 2016).

As noted above, framework analysis (Mason, Mirza, & Webb, 2018) was used to identify the range and salience of key items and concepts, and

Figure 2. The three change dimensions of people, process and technology in ERP systems projects
Source: Wynn and Rezaeian (2015, p. 87)

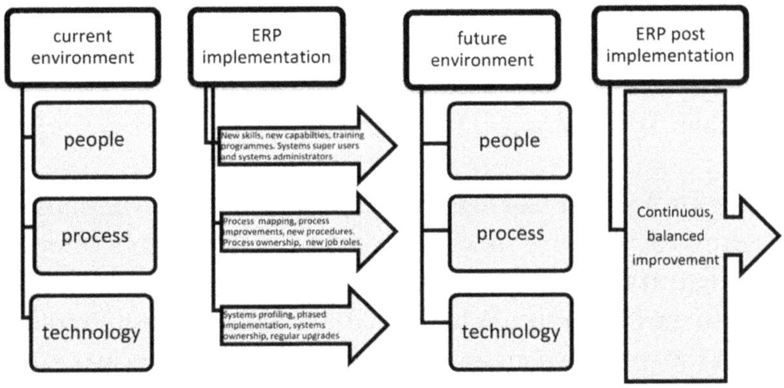

to discover relationships between them. In inductively analyzing the main documentary sources and personal observation notes, a two dimensional framework was constructed comprising the three dimensions of change (people, process and technology) on one axis, and the nineteen projects on the other axis. Building upon an initial pilot research study of three projects (Wynn, 2018), the sources were searched for data relevant to any particular cell in this framework. This produced a matrix of thematic content assigned to the three areas of operation. The data was processed from the matrix format into a mind map for each project, which enabled a further clustering of themes and related issues. This structurally coded data was then synthesised into twelve main change factors that were of relevance to all projects (Figure 3). These are:

Technology Dimension

1. **Requirements Specification**
   a. **Functional Specification:** Was there a clear statement of what the technology should do from a user/functional perspective?
   b. **Technical Specification:** Was there a clear statement of the technical aspects that the new technology must adhere to (database, programming language, operating system, technology strategy)?
   c. **Sign-Off Process:** Was a clear sign-off process for computer users and technical support staff in place and adhered to?
2. **Product Selection and Fit**

a. **Packaged Software:** How well did the system meet the needs of the company? Did it have the required functionality? Was customization necessary? Was it user-friendly? Was it generally accepted by the end-user community?
b. **Development Tools:** If software was developed as part of the project, how appropriate were the tools used?
c. **Other Technologies:** If other technologies were introduced, were they a good fit to requirements?
d. **Technology Performance:** Were there any performance issues in terms of response times, overheads on processors or networks?

3. **Implementation Execution**
   a. **Project Planning:** What was the project planning process? What did the project plan look like and how effective was it in guiding the implementation process?
   b. **Resourcing:** Were adequate human and financial resources allocated to project implementation? Were computer users made available for systems testing and training?
   c. **Product Delivery:** Were the main products delivered on time and to specification?
   d. **Contingencies:** Were contingencies needed to complete implementation and were they made available?
4. **Technology Absorption and Handover**
   a. **Technical Documentation:** Were new products adequately documented, with appropriate walkthroughs of material with the company technical support staff?
   b. **User Manuals:** Were user manuals made available, used in training and formally handed over?
   c. **End-User Champions:** Were key users identified to take on on-going maintenance and support.

Process Dimension

1. **Project Alignment to Business Strategy**
   a. **Clarity and Communication:** Was the company's business strategy clearly documented and communicated?
   b. **Project Alignment:** Was the project clearly aligned to the business strategy? Was the business strategy referenced in the project submission documentation?

Figure 3. The 12 change factors underpinning successful technology transfer

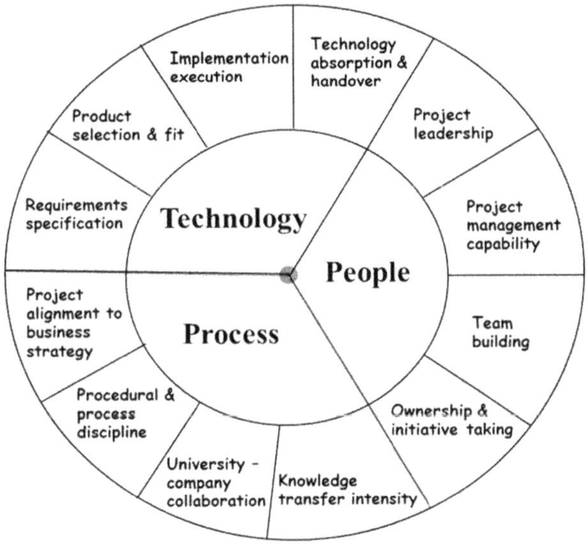

c. **Strategy Evolution:** Were project priorities reviewed in the light of changes in business strategy?

2. **Procedural and Process Discipline**
   a. **Adherence to Agreed Meeting Format:** Did weekly project review meetings take place? Did the quarterly Project Board meetings take place? Were these meetings adequately attended? Were they minuted and were members held to account regarding agreed actions?
   b. **Evidence of Project Management Deliverables:** Were appropriate project management reports, briefs, quality plans etc. developed, signed off and made available for consultation and review?
   c. **Documentation Control:** Were all key documents efficiently stored, version controlled and made available to appropriate personnel.
3. **University-Company Collaboration**
   a. **Academic Supervisor/Lead Roles:** Did the academic supervisor/ lead contribute to the formulation of the project proposal? What role did he/she play in implementation?
   b. **Wider Collaborative Actions:** What other university-company activities paralleled or followed on from the project (e.g. follow-on KTPs, university delivered training)?
4. **Knowledge Transfer Intensity**

a. **Audit:** By project end, what were the main categories of knowledge transferred in terms of technical knowledge and skills, project management capability, process change concepts etc.? What was the depth and significance of this knowledge regarding the project?
b. **Post-Project Review:** Did the company retain the skills and knowledge adequately to continue support and progression of new technologies implemented in the project?

People Dimension

1. **Project Leadership**
   a. **Role of Company Project Sponsor:** Did the project sponsor play an effective role in managing the project and resolving issues within the company (e.g. of resourcing, inter-personnel problems, project planning issues). Was there an escalation process in place to get issues recognized and resolved at the senior management level?
   b. **Role of Academic Team (Associate and Supervisor):** Did the Associate play a significant role in project leadership within the company? Was he/she viewed as the project manager by in-house staff? Did he/she lead effectively, in terms of communication and relationship building? Did the academic supervisor play an effective support and facilitating role, and generally contribute to problem resolution?
2. **Project Management Capability**
   a. **Project Management Knowledge and Training:** To what extent was project management knowledge in evidence? In particular, had the Associate project management experience? Was project management training undertaken prior to, or during, the project for key members of the project team?
   b. **Use of Project Management Methodology:** Was a formal project management methodology used to manage the project? If not, what less formal methods were deployed?
   c. **Documentation:** What project management documents were generated and stored as part of the project? Were they signed off by appropriate staff? Were they made readily accessible? Were they seen as key documents and used effectively in the project?
3. **Team Building**
   a. **Team Formation:** Was a project team formally established to run the project under the direction of a team leader? What was the

Figure 4. Two-dimensional model for assessing focus of technology transfer and degree of innovation

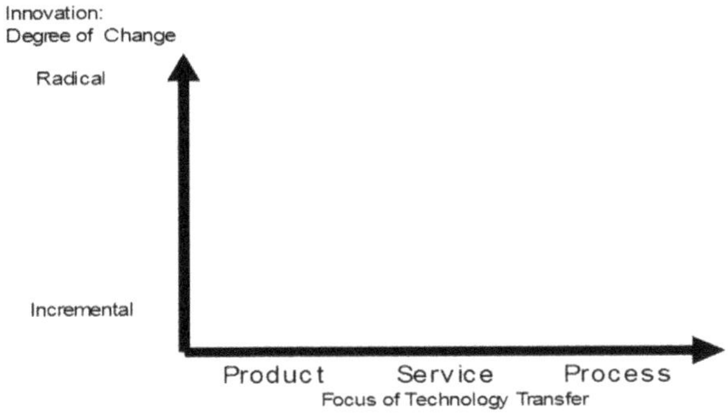

composition of the team – did it contain appropriate personnel from the university and the company departments affected by the project?

b. **Team Operation:** How did the team operate? Were there formal regular meetings? Did it rely on informal communication? Were actions agreed and minuted?

c. **Team Performance:** Was the project team effective in carrying out the key project planning and implementation tasks? Did the team have the combination of skills, knowledge and competencies to perform effectively? Did the team work well together?

4. **Ownership and Initiative Taking**

a. **Company Project Ownership:** Was the project well sponsored and supported by senior management in the company? Was there a clear route for issue escalation and resolution when problems arose?

b. **Associate and Supervisor Ownership:** Were the university staff fully engaged in project delivery? Did they align adequately with company culture and values?

c. **Initiative Impact:** Was there evidence of new initiatives to promote the technology transfer and project delivery processes? Did the Associate demonstrate the knowledge, capability and confidence to make an impact with new ideas and initiatives? Was this in evidence from the academic supervisor or other project team members?

The resultant model will help fill the gap in the existing literature identified by Filippetti and Savona (2017) when referring to university-industry linkages.

They suggest that "it is widely recognized that universities and other public research institutions play a central role within systems of innovation for basic research generation, technology transfer and knowledge diffusion to firms" (p.720), but nevertheless "the analyses of factors that slow down or hamper cooperation have been rather overlooked" (p.724). The technology transfer projects examined in this book are from a range of industry sectors, and, prior to project commencement, the companies' use of technology was generally at a low to medium level, as defined by Alvarez and Iske (2015). As noted above in the Preface, several authors (Urabe, 1988; Popadiuk & Choo, 2006) have seen innovation as being applied to either a product, process or service, and Puerta Sierra and Jasso Villazul (2018) extend this to technology transfer, considering it to be "a voluntary and active process of disseminating or acquiring new experiences or knowledge, with the purpose of improving products, services and productive processes of enterprises" (p.2). This concept is adopted in the assessment of the technology transfer projects studied here, with each project focusing, in the main, on either process change, service improvement or product development. Similarly, the distinction between radical and incremental innovation (Pedersen & Dalum, 2004) discussed above is also incorporated into a two-dimensional model which is used for comparison of projects in each industry sector (Figure 4).

REFERENCES

Alvarez, H., & Iske, P. (2015). Internal Capabilities and External Knowledge Sourcing for Product Innovation in LMT SMEs. *Journal of Innovation Management*, *3*(2), 55–70.

Biggar Economics. (2015). *Economic impact of the University of Gloucestershire*. Retrieved September 12th, 2017 from http://www.glos.ac.uk/docs/download/Business/economic-impact-of-the-university-of-gloucestershire.pdf

Cunningham, J. A., Menter, M., & Young, C. (2017). A review of qualitative case methods trends and themes used in technology transfer research. *The Journal of Technology Transfer*, *42*(4), 923–956. doi:10.100710961-016-9491-6

Department of Trade and Industry. (2003). *DTI Innovation Report: Competing in the global economy: the innovation challenge*. Retrieved October 18th, 2014 from http://www.saiie.co.za/ocs/public/conferences/1/schedConfs/1/program-en_US.pdf#page=46

Department of Trade and Industry. (2004). *Business in the information age: International Benchmarking Study 2004*. London: Booz Allen Hamilton.

Enkel, E., Bell, J., & Hogenkamp, H. (2011). Open Innovation Maturity Framework. *International Journal of Innovation Management*, *15*(6), 1161–1189. doi:10.1142/S1363919611003696

Essman, H., & Du Preez, N. (2009). *Practical cases of assessing innovation capability with a theoretical model: The process and findings*. Retrieved March 12th, 2016 from http://www.saiie.co.za/ocs/public/conferences/1/schedConfs/1/program-en_US.pdf#page=46

Filippetti, A., & Savona, M. J. (2017). University–industry linkages and academic engagements: Individual behaviours and firms' barriers. Introduction to the special section. *The Journal of Technology Transfer*, *42*(4), 719–729. doi:10.100710961-017-9576-x

Gale, N., Health, G., Cameron, E., Rashid, S., & Redwood, S. (2013). Using the framework method for the analysis of qualitative data in multi-disciplinary health research. *BMC Medical Research Methodology*, *13*(1), 117. doi:10.1186/1471-2288-13-117 PMID:24047204

Hannabuss, S. (2001). A wider view of knowledge. *Library Management*, *22*(8), 357–363. doi:10.1108/01435120110406273

Hartley, J. (2004). *Case Study Research*. SAGE Publications. doi:10.4135/9781446280119.n26

King-Turner, M. (2014). *Secrets of successful ERP/CRM projects*. Training Course Handbook. Warwick: The National B2B Centre Ltd.

Knockaert, M., Ucbasaran, D., Wright, M., & Clarysse, B. (2011). The Relationship between Knowledge Transfer, Top Management Team Composition, and Performance: The Case of Science-Based Entrepreneurial Firms. *Entrepreneurship Theory and Practice*, *35*(4), 777–803. doi:10.1111/j.1540-6520.2010.00405.x

Levy, M., & Powell, P. (2003). SME Internet Adoption: Towards a Transporter Model. *International Journal of Electronic Commerce and Business Media, 13*, 173–181.

Lipscomb, M., & McEwan, A. M. (2001). The TCS model: An effective method of technology transfer at Kingston University, UK. *Industry and Higher Education, 15*(December), 393–401. doi:10.5367/000000001101295984

Mason, W., Mirza, N., & Webb, C. (2018). Using the Framework Method to Analyze Mixed-Methods Case studies. *Sage Research Methods Cases, Part 2*. DOI: . doi:10.4135/9781526438683

Neale, B. (2012). Qualitative Longitudinal Research: An introduction to the Timescapes Methods Guides series. *Timescapes Methods Guides Series 1*. Retrieved June 18th, 2018 from http://followingfathers.leeds.ac.uk/findings-and-publications/

Nolan, R. L. (1979). Managing the crisis in data processing. *Harvard Business Review, 57*(2).

Parkinson, S., Eatough, V., Holmes, J., Stapley, E., & Midgley, N. (2016). Framework analysis: A worked example of a study exploring young people's experiences of depression. *Qualitative Research in Psychology, 13*(2), 109–129. doi:10.1080/14780887.2015.1119228

Pedersen, C., & Dalum, B. (2004). *Incremental versus radical change - the case of the digital north Denmark Program*. Paper presented at the 10th International Schumpeter Society Conference, Milano, Italy.

Popadiuk, S., & Choo, C. W. (2006). Innovation and Knowledge Creation – How are these concepts related? *International Journal of Information Management, 26*(4), 302–312. doi:10.1016/j.ijinfomgt.2006.03.011

Puerta Sierra, L., & Jasso Villazul, S. (2018). University-industry linkage and technology transfer management: A quantitative study of an entrepreneurial university. *Journal of Management Development*. doi:10.1108/JMD-06-2017-0203

Ritchie, J., & Spencer, L. (1994). Qualitative data analysis for applied policy research. In A. Bryman & R. Burgess (Eds.), *Analyzing Qualitative Data* (pp. 173-194). London: Taylor & Frances.

Saunders, M., Lewis, P., & Thornhill, A. (2009). *Research Methods for Business Students* (5th ed.). Pearson Education Limited.

Serenko, A., Bontis, N., & Hull, E. (2011). Practical relevance of knowledge management and intellectual capital scholarly research: Books as knowledge translation agents. *Knowledge and Process Management*, *18*(1), 1–9. doi:10.1002/kpm.363

Silverman, D. (2013). *Doing Qualitative Research*. SAGE Publications.

Simmons, P. G., Dawley, D., Richie, W., & Anthony, W. (2001). An exploratory examination of the knowledge transfer of strategic management concepts from the academic environment to practicing managers. *Journal of Managerial Issues*, *13*(3), 360–376.

Taylor, M., & Murphy, A. (2004). SMEs and the take-up of E-business. *Urban Geography*, *25*(4), 315–331. doi:10.2747/0272-3638.25.4.315

Technology Strategy Board. (2013). *Concept to Commercialisation: A strategy for business innovation, 2011-2015*. Swindon: Technology Strategy Board. Retrieved September 24th, 2018 from https://assets.publishing.service.gov.uk/government/uploads/system/uploads/attachment_data/file/360620/Concept_to_Commercialisation_-_A_Strategy_for_Business_Innovation_2011-2015.pdf

Urabe, K. (1988). Innovation and the Japanese Management System. In K. Urabe, J. Child, & T. Kagono (Eds.), *Innovation and management international comparisons*. Berlin: Walter de Gruyter. doi:10.1515/9783110864519-005

Walsham, G. (1995). Interpretive case studies in IS research: Nature and method. *European Journal of Information Systems*, *4*(2), 74–81. doi:10.1057/ejis.1995.9

Willcocks, L., & Sauer, C. (2000). *Moving to E-Business*. London: Random House.

Wu, W.-L., Hsu, B.-F., & Yeh, R.-S. (2007). Fostering the determinants of knowledge transfer: A team-level analysis. *Journal of Information Science*, *33*(3), 326–339. doi:10.1177/0165551506070733

Wynn, M. (2018). Technology transfer projects in the UK: An analysis of university-industry collaboration. *International Journal of Knowledge Management*, *14*(2), 52–72. doi:10.4018/IJKM.2018040104

Wynn, M., & Jones P. (2017). Knowledge Transfer Partnerships and the entrepreneurial university. *Industry and Higher Education, 31(*4), 267–278. doi:10.1177/0950422217705442

Wynn, M., & Rezaeian, M. (2015). ERP implementation in manufacturing SMEs: Lessons from the Knowledge Transfer Partnership scheme. *InImpact: The Journal of Innovation Impact, 8*(1), 75-92.

Wynn, M., & Turner, P. (2013). Effecting successful knowledge transfer: Lessons from the UK Knowledge Transfer Partnership scheme. *International Journal of Management Education*, *7*(3), 293–312.

Wynn, M., Turner, P., Banik, A., & Duckworth, A. G. (2016). The impact of customer relationship management systems in small businesses. *Strategic Change, 25*(6), 655–670. DOI: doi:10.1002/jsc.2100

Yin, R. K. (2012). *Applications of Case Study Research* (3rd ed.). London: SAGE Publications.

Chapter 1
Contract Packers

ABSTRACT

The rise of demand and supply chain management is of particular relevance to outsourced contracting companies as they are but one cog in a complex chain and have no end products of their own. They rely upon a range of customers and suppliers (the two sometimes being the same entity), who are in turn dependent upon the contract packer. This leads to a complex supply chain where profit margins are generally small and where internal process efficiency and excellent customer service are critical for survival. In this chapter, three projects that introduced new technologies to improve processes or services are reviewed. They all aimed to increase revenues and drive down costs but achieved mixed outcomes.

INTRODUCTION

In the past two decades, the packaging industry has observed large changes in its deployment of technology. Coles (2003) suggests that this "can be of strategic importance to a company, as it can be a key to competitive advantage" which may be achieved through "enabling lower costs, increasing margins, enhancing product/brand differentiation, and improving the logistics services to customers" (p.4). This has given rise to a growth in the contract packers, as some companies decide to outsource their packaging operations and focus on what they consider to be their core business. The industry has its own trade association, the British Contract Manufacturers and Packers Association (BCMPA), which operates throughout the UK and European packaging

DOI: 10.4018/978-1-5225-7408-8.ch001

industry, and provides a valued resource both for members and for brand owners, retailers and manufacturers looking to outsource. It has over 140 accredited members who provide an extensive range of services including contract packing, contract manufacturing, filling, fulfilment, logistics and supply chain management across many industry sectors.

The focus of this chapter is not on the packing technology *per se*, but rather on projects that introduced information and laboratory systems technologies into two contract packing companies, Brecon Pharmaceuticals Ltd and Contrapak Ltd.

NEW INTEGRATED INFORMATION SYSTEMS AT BRECON PHARMACEUTICALS, HAY-ON-WYE

Company Profile

In 2003, Brecon Pharmaceuticals Ltd (BPL) was an independent company based in the Wye Valley Business Park, near Hay-on-Wye in Hereford, UK. The company provided "outsourced" services to the pharmaceuticals and healthcare industries, principally the packaging of manufactured drugs and of new drugs undergoing clinical trials. BPL's reputation was founded upon a high quality, flexible service. The company's corporate vision was to achieve "growth through excellence" where "excellence" is defined as "the customers' confidence in the company's ability to deliver according to their requirements" (Momenta, 2005, p.2). The market for outsourcing in the UK pharmaceuticals industry was circa £50m per annum and was expanding by at least 5% per year. BPL had a 10% share - its main competitors being the other UK outsourced packaging providers, including Unipack (£20m turnover) and Penn Pharmaceuticals Ltd (£14m turnover). BPL's business plan was to grow turnover from £4m in financial year 2001 to £22m in 2007, increasing pre-tax profits from £0.5m to £4.4m over this period. By 2005/6, however, the company had increased its turnover to £11m, and with substantial investment in new packaging facilities and successful contracts agreed with international drug companies (e.g. Eli Lilly), was well placed to pursue its growth ambitions. Today, in 2018, after a number of sales and mergers, the company is now part of Packaging Coordinators Inc (PCI), a global packaging business owned by private equity company Frazier Healthcare.

Project Rationale

The ability to respond rapidly to customer requirements is a key factor in obtaining business and BPL held some spare capacity, in terms of machinery and facilities, which could be "switched on" at short notice. Typically, the firm order book was only two to three months long, but it took approximately six months to recruit and train staff to the required standard or, where necessary, to order and install new equipment. In order to maintain its competitive advantage, BPL had to optimise the management of its resources, making efficient use of existing facilities and timing the introduction of new equipment and qualified staff to best effect.

As the company grew, it became clear that its information systems needed to be robust and scalable to support the planned increases in business volumes. In line with this growth, the number of employees increased from 162 in 2003 to over 300 in 2006, bringing associated challenges in managing employee data, and the administrative burden of an increasing range of employee benefits. BPL had to replace the existing mix of manual and non-integrated computerised systems, which were inadequate for an expanding business and could have seriously undermined the planned expansion. It was also accepted that these new systems would need to be accompanied by significant process change, but the company did not have the in-house skills necessary to introduce technology and process change simultaneously on such a scale.

Project Overview

The 2 year project plan (Figure 1) comprised three distinct systems sub-projects. The first, scheduled to last 15 months, was to implement a new Enterprise Resource Planning (ERP) system. With this in place, a separate module for human resource management would be acquired and installed, and then a company Intranet was to be implemented. However, when the project got underway in 2003, it became clear that the particular requirements of the pharmaceuticals industry in relation to Good Pharmaceutical Manufacturing Practice (GPMP), stock traceability, and process validation, required that the roles that each department played in delivering their services to clients needed to be assessed and documented prior to software selection. This understanding was key in ensuring an appropriate package was selected that would support the company's business processes. The Associate took charge of the organisation of workshops and process analysis sessions that formed

Figure 1. The outline project plan for ERP, human resource management and Intranet systems at BPL (Teaching Company Directorate, 2003, p.19)

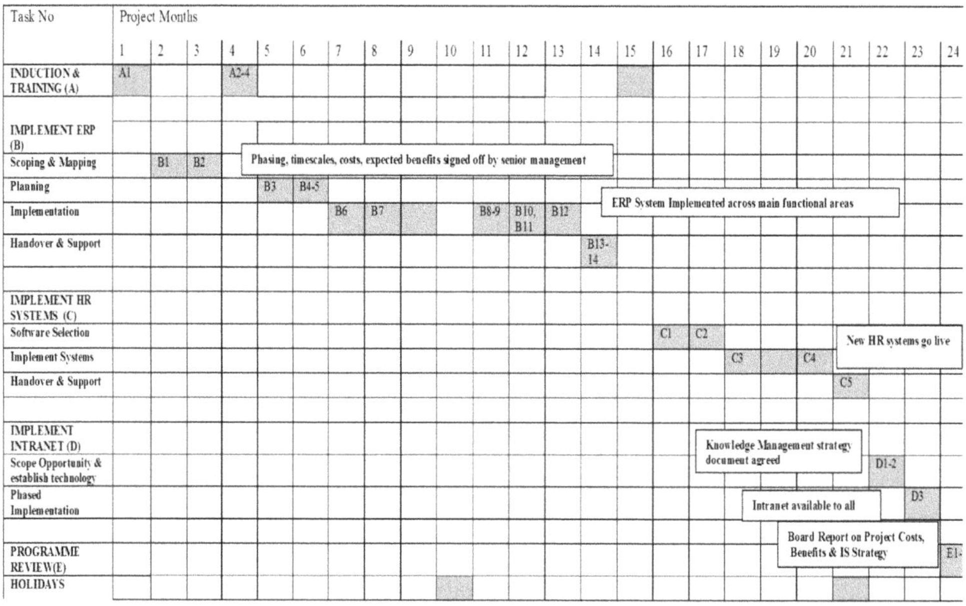

the basis of the software requirements process. He set up and integrated the project team effectively and diplomatically within a very short time frame. This culminated in the selection of the EFACS ERP software package as the core system for the company's future information systems requirements. It was judged by the project team as adequately meeting most critical requirements, with the exception of the payroll and human resource management functionality that was to be addressed by the second sub-project, and for the while would continue to be supported by the old legacy systems. The identified need for a company Intranet was also seen as secondary, to be addressed after ERP implementation. This allowed the project team to focus on the main ERP implementation.

Implementation Issues

During the implementation of the EFACS software, a number of unexpected problems surfaced. There were delays in the sign-off of the validation of compliance of the software with GPMP and stock traceability requirements. Due to lack of experience of implementing software within the company, new procedures had to be developed, tested and approved before the application could be signed off as fit for use. The issue of validation was further complicated by the need to comply with pharmaceutical regulations dictated by the European Union's Guidance on Manufacture. Then, when the new ERP system was finally migrated from a test environment to the "live" environment, it was found that several of the client PCs could not gain access to the server. This was because the client PCs had been set up to operate on a closed network, and these settings had been hard coded in the registry files of particular client PCs. This issue further delayed the validation of the ERP software, as the cause of the problem was not one that had been experienced in the past by consultants from the software supplier. In addition, a number of bugs were found in the ERP system, which required either code changes or updates to be received from the supplier. Having pursued the occurrence of these bugs, the software supplier responded that constant changes were being made to enhance the application as new versions were released. Despite efforts to release a bug free application, some bugs were missed.

In parallel with these implementation delays, priorities regarding the secondary projects included in the original project plan had to be changed. Replacement of the pre-existing systems for document management and label production was not within scope of the original project, and yet these systems were also governed by GPMP standards and, as such, were subject to frequent audits by clients. The document management system required an urgent upgrade and there were a number of validation and support issues with the label production software that needed attention. It was thus decided that these should be given priority, and the HR/payroll and Intranet projects were therefore put on hold. It was thus no longer possible to achieve the original project objective of implementing a new HR/payroll system and Intranet alongside the ERP system. However, the Associate was able to deliver the document management system upgrade and he also took on more responsibility relating to the general management of IT activities. This proved of value after the close of the project as the Associate became an employee of the company as their IT manager.

Project Benefits

Despite the issues noted above, there were nevertheless some very significant project benefits. In the course of installing the EFACS ERP system, the company learned a great deal about the process of implementing a major new software system in many parts of the company simultaneously, and the amount of resource needed to achieve a successful installation. BPL also reviewed existing processes and, in many cases, was able to modify or refine them in order to maximise the benefits from major systems change.

The ERP and document management systems “improved the company’s ability to manage the increasing variety of client orders” (Momenta, 2005, p.3), allowing the company to routinely process over 60 works orders every month. Each works order required a significant amount of supporting documentation, and materials, machines and staff needed to be available at the right time, in the right configuration, in order to maximise outputs and efficiency. The Final Report (Momenta, 2005) noted that “this level of activity could not have been sustained without the new IT capability” (p.3). Staff awareness of the demands of running an integrated software product had been raised significantly, and the IT literacy of staff in general had increased as a result. IT-based solutions became the norm, representing a culture shift in the company’s approach to problem solving.

The company’s ability to forward plan was also much improved, allowing important investment decisions to be made with the best possible supporting information. It was important, as a service provider, that the company did not make delivery promises that could not be achieved, thereby undermining client confidence. Planning capabilities were extremely important and none of the company’s competitors had the level of sophistication for capacity planning brought in with the new ERP system. Specific quantifiable benefits included headcount savings in production, quality assurance, warehousing and finance, delivering an £80K saving per annum (4 staff). The increase in cubicle utilisation brought about by improved scheduling and production planning was quantified as a hard benefit of £640K per annum (Momenta, 2005, p.4).

Analysis

This project was given a B rating by Momenta (the UK Government funding body), and the Final Reports indicate a general satisfaction with the project outcomes by all parties involved (Figure 2). Key to the success of the project was the strong leadership provided by the finance director who was the project sponsor and chaired the weekly project review meetings, and the clear alignment of the project with the company’s growth strategy, which was clearly communicated and cascaded down through the organisation. The finance director gave the project a direct route into the managing director that

Figure 2. Change factors in the BPL information systems project

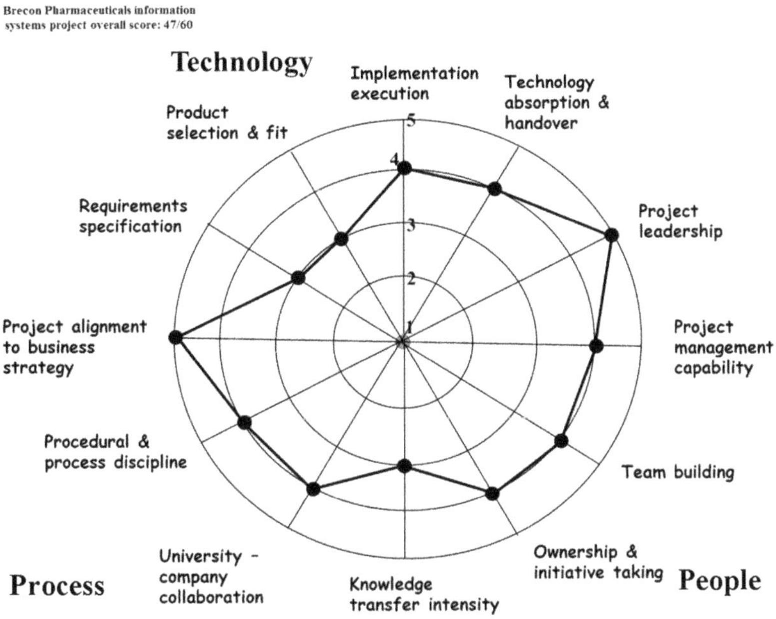

was particularly significant in resolving prioritisation and resourcing issues, and coordinating activities in different business areas.

The Associate was a qualified project manager who instilled a clear discipline and procedure into the day to day management of the project, and the project team worked well in taking ownership of, and resolving, a range of operational issues. The requirements specification process had actually started prior to the formal involvement of the University, and this could arguably have been done better. There was no clear understanding of the implications of GPMP at the outset, and this caused some delay in the early months of the project. Eventually an ERP package was selected that suited the procedural requirements of GPMP and stock traceability, but which did not have human resource management and payroll functionality. The software was also not bug-free, and this contributed to implementation problems requiring the allocation of extra resources. This was not a critical weakness, but impacted timescales and overall project outcomes.

In summary, the project was weakest in the initial activities of requirements specification and product selection, which required the scheduling of additional sub-projects for human resources management, payroll and Intranet systems - projects which did not happen within the two year project timeframe.

The main strengths lay in the leadership provided by the company finance director and the Associate, the alignment of the project with overall business strategy, and the positive attitude and approach of the staff involved – team building, ownership and initiative taking, and project management capability all contributed positively to a successful implementation of the ERP system.

SUPPLY CHAIN INTEGRATION AT CONTRAPAK, HEREFORD

Company Profile

Contrapak Ltd was a small contract packaging business located on two sites near to Hereford, with their own dedicated production facilities based at Kingstone (which primarily progressed orders related to contract packing of chemicals) and Rotherwas (which produced liquid and hand packing solutions). In 2009, when the technology transfer project started, the company employed 52 staff, and had a turnover of £1.2m. The company's areas of expertise were sachets and bagging, blending and formulation, shrink wrapping, labeling, pouch filling, liquid filling, high speed counting, hand assembly and repacking. The key strengths of the business at the time were seen as chemical powder blending and the ability to fill into virtually any container or sachet/bag available on the market. Contrapak also had the largest dedicated water soluble packing unit in the country.

Project Rationale

Contrapak's business processes were only partly automated in 2009 when they developed their business plan for the ensuing 5-year period. They had acquired software packages for financial management and payroll, but the key process areas of production planning and stock control remained largely manual. The company's senior management concluded that new systems were required to support the significant increases in turnover targeted in the 5-year plan. In particular, the lack of integrated systems was causing major problems in effectively scheduling customer jobs on the various packaging lines, and there was a clear need to reduce inventories and improve production lead times. There was a dire lack of consistent and accurate management information and many of the key processes were "silo-like" and not properly integrated.

As the project progressed, the need to integrate planning with upstream order capture information came to the fore and led to the adoption of a more holistic view of the supply chain process in the selection of a new system.

Requirements Specification and Package Selection

In late 2009, the company embarked on a systems upgrade project with the University of Gloucestershire through the Knowledge Transfer Partnership (KTP) scheme. The Associate interviewed staff across the company to develop a top-line process map and profile existing systems, as well as identifying opportunities for process improvement. The project operated through a Project Board (comprising the five senior managers and the university supervisor and Associate), and a project team (consisting of the Associate, production operations manager, and key operatives). Analysis of current business needs was carried out, focusing in particular on the planning and management of the packaging lines and related stock control. Detailed flow charts for current business processes were mapped graphically to facilitate a better understanding of these processes and the role of supporting documents. Key personnel were involved in workshops to verify process flow details (Figure 3) and establish current and future information needs. The findings

Figure 3. Process flow in sales and production areas at Contrapak 2010 (Wynn & Olubanjo, 2012, p.202)

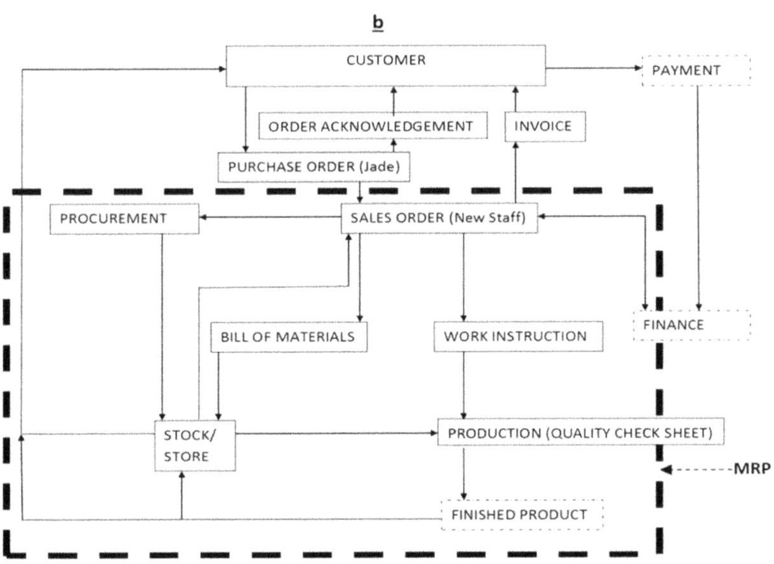

were used as the basis for specifying the critical improvements that new systems needed to deliver. Using a red-amber-green analysis, the strategic soundness and capabilities of existing systems were mapped in each process area (Figure 4). The project team also spent some time focusing on key data items, particularly customer and product data, and establishing where these would be entered and processed in new systems. From this range of new information, the project team developed a system requirements specification document for all main process areas.

The requirements specification document was sent out to the prospective vendors, and five suppliers were identified that could provide the required functionality in the packaging and stock control areas of the business at acceptable cost. Four of these software houses also offered other modules covering some of the other main business process areas – order capture and customer relationship management, finance and ledgers, plant maintenance and quality control. This was to prove instrumental in the change of scope that followed. Other factors that were considered included assessment of how the software solution would perform on Contrapak's current IT infrastructure – network capability, operating system, database requirements.

The five software companies presented their products at Contrapak over three separate days in 2011. The full Project Board, comprising managing

Figure 4. Process mapping and systems profiling at Contrapak (Wynn & Olubanjo, 2012, p.205). Systems status 2009: R = Red – replace, A = Amber – possibly retain, G = Green – can be retained

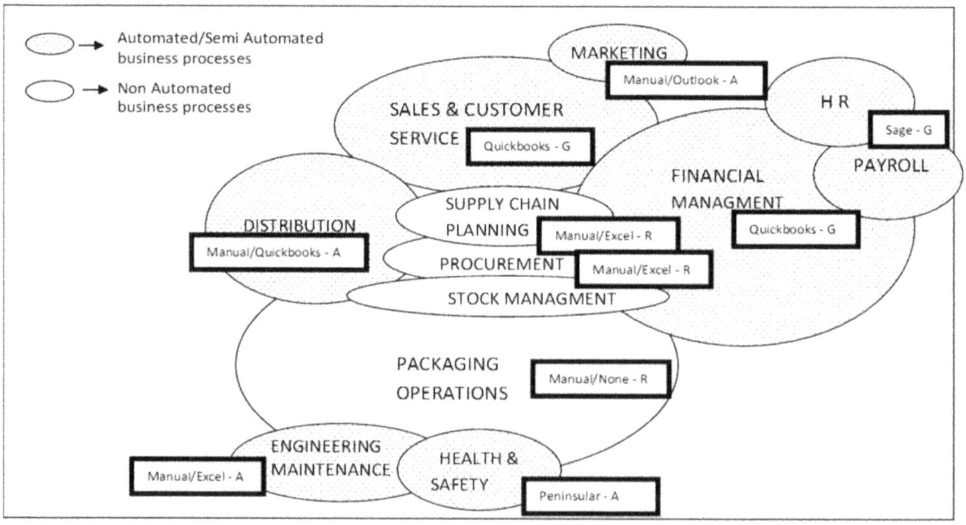

director, finance director, sales and marketing manager, production operations manager and technical director, plus the Associate and supervisor from the university were present. The five products on the short list were Factory Master, Sage, Priority EMS, Infor and EFACS, and they were scored against 9 criteria. The company entered into commercial negotiations with the two companies scoring highest, and after considerable debate, elected to implement the Infor ERP product.

Project Implementation

After a week of in-depth testing, the Project Board agreed to expand the project scope to include new software modules for order entry and financial ledgers from Infor to complement the production planning and stock control modules from the same supplier. This recognized the integration of supply and demand chain processes and necessitated a re-planning of the project. This project had initially been scheduled to run for just 9 months (a "short" KTP), but with the expansion of project scope the company and funding body agreed to a further 9 months support, and senior management also signed off a limited number of days consultancy from Infor to help set up and configure the software modules. The re-worked project plan (Figure 5) focused first of all on setting up the basic product information in the master item file, followed by the transfer and cleansing of all the basic financial data into the new ledgers, thereby allowing the turning off of the old QuickBooks legacy finance system.

With the new financials module in place, a second phase focused on implementing the new order entry functions in the Infor system. However, at

Figure 5. The revised plan for the Contrapak systems project (Wynn & Olubanjo, 2012, p.207)

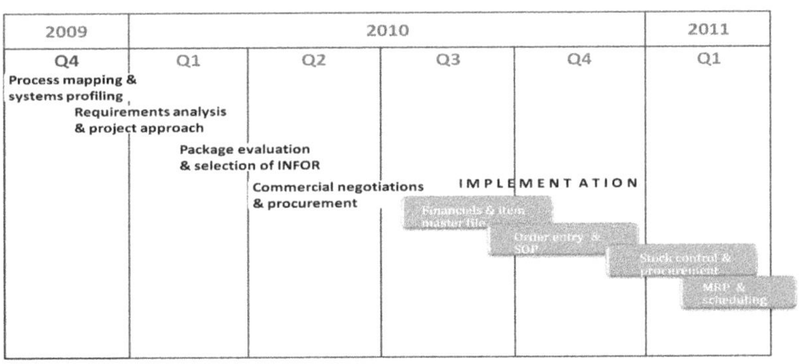

this point, the project began to slip due to a lack of user resource and input into a number of essential activities – the setting up of customer data and bill of materials data for packaged products, the configuration of the new systems modules for stock control and material requirements planning (MRP), and user testing. In early 2011, after an 18 month extended timeline, the involvement of the university ended. The project manager role was handed over to the production operations manager, whose first task was to progress the testing and commissioning of the MRP module and related stock functions of the software suite. These functions were to allow automated materials planning, producing the scheduling and control of product supply and necessary inventory movements, but again progress was slowed by resource problems – staff were not available to undertake these tasks, and there was no budget available for more consultancy support from the software vendor. Finally, in 2012, the Infor ERP package went live, and the business functions of financials, order entry, planning and stock control were managed, albeit briefly and with difficulty, via the new system.

Project Outcomes

The running of the new Infor system had many teething problems, but the company was starting to see the potential benefits of packaging machine use optimization and integration of the planning and scheduling functions, when in 2013, the company went into administration. The company had over-reached itself in attempting to undertake a large contract with a new high profile customer, and the company's liabilities became unmanageable. However, the vast majority of the packaging lines and labeling facilities were sold on to a new entity, Wyepak Ltd (managed by the former technical director and production operations manager from Contrapak), who are now trading nearby in a new 20,000 sq. ft. factory in the Rotherwas enterprise zone in Hereford.

The burden of implementing the new ERP system was not of major significance in the collapse of the company, but with hindsight it is clear that financial constraints in this low margin business limited the amount of human and other resources that were available at the critical phase of systems implementation, when the in-house staff took on full responsibilities for managing the project. With the problems associated with the imminent closure of the company in 2013, the ERP project lost momentum, and the Infor system is not currently used by Wyepak, in part because of issues of license transfer between the two companies and on-going maintenance costs.

Wyepak are using cloud based software for accounts and sales and marketing, and Microsoft Excel spreadsheets for stock management and production scheduling.

Analysis

The Contrapak project was officially a "short" 9 month KTP (albeit it was granted a further 9 months extension), and was thus not formally graded by the UK Government's funding agency. Nevertheless, the change factors diagram (Figure 6) reflects the initial fast and successful start to the project, which regrettably then fell away as the company got into financial difficulties. The project was well conceived and initially led from the front by the managing director and his senior team. They realized that the company needed to automate and integrate its core sales order processing and packaging functions, and - well supported and facilitated by the Associate - requirements specification and software selection were thoroughly and intensely pursued, with full commitment from the Project Board and project team. Exposure to the realities of implementing a major supply chain project provided those involved in the project with knowledge of the ERP concept and an understanding of the

Figure 6. Change factors in the Contrapak ERP project

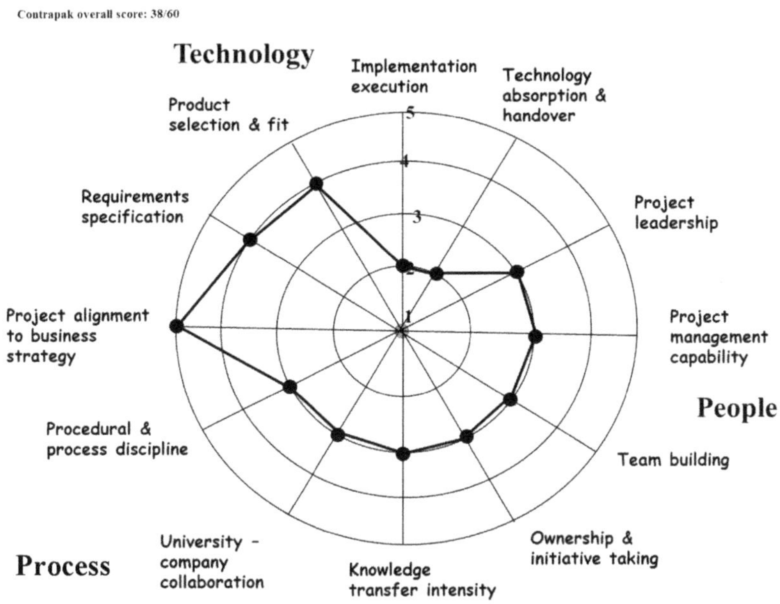

resources required to get a major software product successfully implemented. They also had some exposure to elements of the PRINCE2® and PRISM Buy-Build methodologies employed during the project.

Knowledge transfer intensity, team building, ownership and initiative taking, and university-company collaboration were all very positive in the first year of the project, but then diminished rapidly. This resulted in a poor implementation of the new technology and only a partial handover of the system to Contrapak staff. The implementation process was undermined by a lack of focus, and particularly the absence of a full time project manager once the university input ceased. The inability to address process change and systems problems without input from the software vendor consultants severely hampered progress, and as the company's financial problems intensified, the project was put on hold. Technology handover was never properly completed, and the project was essentially left "half- finished". Many of the change factors were initially very positive, but deteriorated over the course of the project. This has resulted in average assessments for many of the change factors.

This was a disappointing outcome to the project, especially as the new system and associated process change could have brought significant benefits from an improved ability to respond to customer requirements - in maintaining existing relationships and securing new business; and also in reducing the overhead on management in resolving operational issues and data conflicts in the packaging and customer facing areas, and in generally improved information quality that could have engendered more effective decision-making. However, pressure from customers - for shorter lead times, for more frequent deliveries, and for specific customer requirements – was on the increase and contributed to the company's financial and operational problems just as the new system was being implemented.

IN-SOURCING LABORATORY SERVICES AT BRECON PHARMACEUTICALS, HAY-ON-WYE

Background

The first case study in this chapter discussed the new information systems project at Brecon Pharmaceuticals Ltd, which provided "outsourced" services to the pharmaceuticals and healthcare industries. These services were twofold - the packaging of manufactured drugs and the testing of new drugs undergoing clinical trials. This case study now assesses a second, very different technology transfer project undertaken at the company. BPL's reputation was founded upon a high quality, flexible service, and in 2004, when this second project started, the company had, itself, contracted out their laboratory testing services to third parties. However, to support their growth plans, the company took the view that clients would in future prefer to place both their packaging and analytical needs with one company, and thus bring new business to the company that would otherwise have gone elsewhere.

At the start of the project in 2004, the company had 231 staff and a turnover of £7.6m, but was targeting a tripling of revenues to £22m by 2007. The technology transfer was to support a major change in service delivery, achieved through the installation of new laboratory hardware and support infrastructure.

Project Objectives

The principal overarching project objective was to establish a state of the art analytical laboratory to undertake routine analysis and associated testing for the company's current and potential client base. The company wanted to develop a center of excellence for laboratory information, statistical analysis and analytical testing, involving a new laboratory information management system (LIMS), which would be linked to their newly installed corporate ERP system. The project would research client requirements for analytical services and develop a new revenue stream from additional analytical work, targeting a £500K per annum revenue increase by 2008. The project focused

Figure 7. The outline project plan for the BPL laboratory services project (Momenta, 2004, p.39)

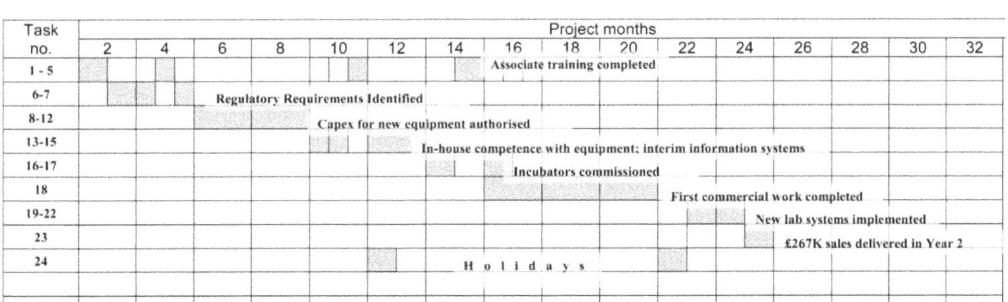

on the introduction of new laboratory equipment and associated systems to provide in-house testing for its clients. As such it used new technology to innovate its service offering in this field.

Project Implementation

The project team were assembled in early 2004, and embarked on a five stage project that was to last 24 months (Figure 7). Stage one involved the research of regulatory requirements and their implications for the client base, and an assessment of the potential new business opportunity. The current and future legislative guidelines for stability and validation testing in the UK, EU and USA were identified and assessed, as was the existing and forthcoming legislation and regulatory frameworks for the clinical trials and laboratory testing sectors. Client requirements for analytical services were also explored at this stage and forecasts made of potential new additional revenue from new clients.

A piloting of new laboratory equipment was then undertaken and relevant configurations and space requirements were identified. Based on pilot results, an equipment short-list was drawn-up with space/facility implications, and procurement of a range of equipment followed. This included high pressure liquid chromatography, ultra violet/infrared scanners, and dissolution, Karl Fisher and disintegration units. The initial set-up and training on new laboratory equipment followed, and for each of the main equipment types, user requirements (as in the specification) were confirmed, installation qualification documents written, and operational qualifications documented. Standard operating procedures for routine use were also drawn up. An interim data storage and analysis tool for analytical data was developed in Microsoft Excel software. Stability incubators, for testing the stability of compounds over time, were installed for use in conjunction with the laboratory equipment.

The new in-house laboratory was ready for commercial packaging and clinical trials operation in 2005. Sampling and testing of clinical and commercial products in accordance with agreed procedures and protocols were undertaken. Laboratory performance and customer responses were reviewed and approaches and techniques refined. An information systems review was undertaken in parallel, leading to the migration of laboratory data into a Microsoft Access database. The project was completed in early 2006.

Project Outcomes

The creation of the laboratory resulted in new client orders coming to BPL. Some clients required analysis prior to importing products into the European Union (EU), whilst others required analytical testing before exporting products outside the EU. BPL rapidly developed links with companies requiring testing for import, export and the internal EU market. Within BPL, the project brought new knowledge and expertise in the deployment of process control equipment, laboratory technologies and test equipment, and project management skills in the development and implementation of these new technologies. The company also improved the scope of assessment systems for chemicals used on site (in compliance with the Control of Substances Hazardous to Health regulations, 2002) and acquired expertise in generating method transfer and stability reports to meet expectations of international clients. The laboratory represented an entirely new function which broadened the services the company could offer. The potential to win more projects was enhanced as the company could offer clients improved flexibility to meet their varying analytical needs. In the broader financial context, BPL continued to grow its turnover from £10.2m in 2004 to £12.9m in 2005. The new laboratory had become a key strategic service to support and promote increases in turnover and profit.

Prior to the start of the project, the company's outsourced laboratory costs were running at £100K per annum, and were projected to rise to over £250K in 2007 as BPL's business grew. The Final Report noted that "not only has this cost been saved, but new additional analytical contract sales are estimated to bring the total revenue gain (saving on outsourced costs plus new sales) to £847K by 2008. The project has generated additional income as a one-off service, competing with the laboratory contract companies BPL had previously used itself" (Technology Strategy Board, 2007, p.3). This new revenue stream generated an additional £151K per annum in financial year 2005 and two significant competing contract analytical laboratories (Wickham and Pharmalytic) ceased to offer an analytical service once the BPL laboratory was in operation. The Final Report concluded that "Brecon is well positioned to benefit from this reduction in competition" (Technology Strategy Board, 2007, p.3).

In March, 2006, the company was bought by the American international pharmaceutical services company AmerisourceBergen, to act as its hub for European operations. The company doubled the size of its operations by expanding the manufacturing, packaging and warehousing facilities in Hay-on-Wye. The expansion was prompted by a growth in the demand for the company's clinical trial supply and commercial packaging services, including the in-house laboratory services. An additional 115 staff were recruited bringing the total staff headcount to more than 300 employees. Then in 2012, AndersonBrecon was created when Brecon Pharmaceuticals merged with US firm Anderson Packaging, and in 2013 AndersonBrecon was acquired by PCI from AmerisourceBergen. PCI now has 16 facilities across North America and Europe, employing over 2,900 highly trained staff, and selling its services in over 100 countries worldwide.

Analysis

This project was given a B rating by the Technology Strategy Board, the UK Government's funding body. Overall, the project was successful in that all major objectives were met. The laboratory represented an entirely new function which broadened the scope and range of services the company could offer. As a result, the potential to win more and varied projects was enhanced. The establishment of the laboratory and associated services supported the company's commitment to offer clients improved flexibility to meet their variable analytical needs and priorities.

Figure 8 indicates the generally positive assessment of all change factors. The only caveat was the emergence of inter-personal issues in the project team towards the end of the project. Although these were managed well, there was some impact on team building effectiveness; but the strength of other aspects of the project - notably the commitment, drive and leadership of the company laboratory manager - ensured the desired project outcomes were delivered. In addition, the alignment of the project to business strategy ensured it was given high profile and the necessary funding for the acquisition of new equipment, which was in line with the requirements which were carefully specified by the Associate and the laboratory manager. Project planning was done effectively using the Microsoft Project tool, involving some minor modifications to the original outline plan. For example, in the original project proposal, it was envisaged that a full LIMS system would be implemented, but after discussion with the IT manager, it was decided to defer this and continue with the interim Microsoft Access and Excel applications, which provided adequate storage, analysis and reporting of laboratory data.

Figure 8. Change factors in the BPL laboratory services project

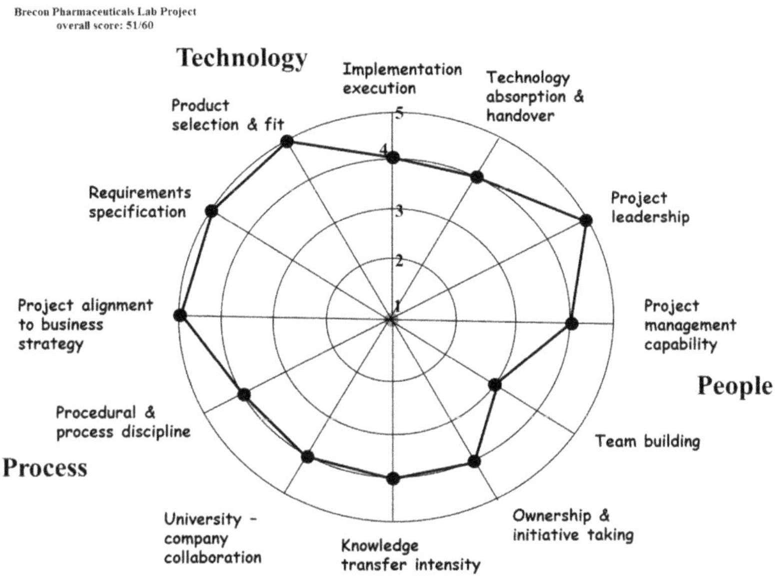

The project was well managed with weekly project review meetings that underpinned an effective handover process and full ownership of the new equipment by company personnel.

CONCLUSION

Supply chain management can be defined as "the management of upstream and downstream relationships with suppliers and customers to deliver superior customer value at less cost to the supply chain as a whole" (Christopher, 2005, p.5). These three projects in the contract packaging industry illustrate how technology was used in different ways to try to achieve this, with improved customer service being a key overall aim, delivered through process improvement in two of the three projects.

The projects also exhibited a number of contrasts. The two projects at BPL were generally successful and contributed to the technology advancement and overall growth of the company. They were of a very different nature in terms of the technology transfer, but both projects were closely aligned to company strategy and strongly led by senior company management. As regards the

Figure 9. Contract packaging company projects: technology transfer focus and degree of innovation

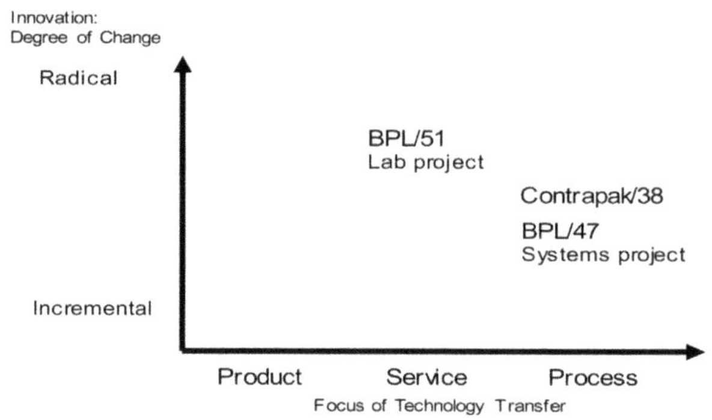

nature of technology transfer, the Contrapak project was similar to the BPL information systems project, in that it attempted to implement a new ERP software package, but was significantly different in terms of resourcing – with limited finances and staff availability ultimately leading to project failure.

The BPL laboratory project was the more radical, in that it attempted and achieved a complete change in the manner in which the company undertook its product testing services, and also aimed to bring in new customers seeking third-party joint packaging and product testing operations. The other two technology projects aimed at process improvement, Contrapak attempting a major change in their supply chain operations (Figure 9).

REFERENCES

Christopher, M. (2005). *Logistics and Supply Chain Management – creating value adding networks* (3rd ed.). Pearson Education Ltd.

Coles, R. (2003). Introduction. In R. Coles, D. McDowell, & M. Kirwan (Eds.), *Food Packaging Technology*. Oxford, UK: Blackwell Publishing Ltd.

Momenta. (2004). *Knowledge Transfer Partnership Project Proposal and Application Form Brecon Pharmaceuticals Lab Project*. TTI Database Record, KTP Programme Office. (Unpublished)

Momenta. (2005). *Knowledge Transfer Partnership Final Report KTP 4291 (Brecon Pharmaceuticals information systems project)*. TTI Database Record, KTP Programme Office. (Unpublished)

Teaching Company Directorate. (2003). *Teaching Company Scheme Grant Application and Proposal Form Brecon Pharmaceuticals*. TCD. (Unpublished)

Technology Strategy Board. (2007). *Knowledge Transfer Partnership Final Report KTP 0167 (Brecon Pharmaceuticals Laboratory Project)*. Database Record. Technology Strategy Board. (Unpublished)

Van Goor, A. R. (2001). *Demand and supply chain management: a logistical challenge*. Paper presented at 17th International Logistics Congress, Thessaloniki, Greece. Retrieved September 12th, 2018 from http://www.sahandtarjomeh.com/wp-content/uploads/2013/07/vangoor-salonique.pdf

Wynn, M., & Olubanjo, O. (2012). Demand-supply chain management: systems implications in an SME packaging business in the UK. *International Journal of Manufacturing Research, 7*(2), 198-212. DOI: doi:10.1504/IJMR.2012.046803

KEY TERMS AND DEFINITIONS

Enterprise Resource Planning (ERP): An integrated software system that supports all business functions in a company. Originally used mainly by large companies, they are now being deployed by small to medium sized enterprises as well. Examples include SAP, Oracle, EFACS, and Sage.

Good Pharmaceutical Manufacturing Practice (GPMP): Is a system and set of procedures that ensures pharmaceutical products are manufactured to established quality standards.

K: Is used to denote thousands, particularly in the context of costs or savings e.g. £150K.

Knowledge Transfer Partnership (KTP): A UK Government scheme to support the transfer of knowledge and expertise from the university sector to companies and organisations. Projects often concern technology transfer and are typically of 2 years duration.

Laboratory Information Management System (LIMS): A software package that is specifically designed to support laboratory operations. Sometimes these systems interface or integrate with the larger corporate ERP system.

Material Requirements Planning (MRP): A software system that is used for production planning, scheduling, and inventory control system in the manufacturing process. Such a system can be one module in a larger ERP system.

PRINCE2 (Projects in Controlled Environments): A mainstream project management methodology used extensively in the UK and internationally by both Government agencies and the private sector. It is strong on quality control and risk management and projects using the methodology are run by a Project Board. ®PRINCE2 is a registered trademark of AXELOS Limited, used under permission of AXELOS Limited.

PRISM (Professional Information Systems Management): A systems development and project management methodology originally developed by Hoskyns. A straightforward, logical, and easy-to-follow approach to systems development and software package acquisition and implementation.

Teaching Company Directorate (TCD): This was the original UK Government agency that ran the Teaching Company Scheme (TCS), the forerunner to the KTP scheme. It was succeeded by Momenta in 2003, then by the Technology Strategy Board in 2007 and Innovate UK in 2014.

Chapter 2
House Builders

ABSTRACT

Larger companies have for some years used customer relationship management (CRM) systems as part of a strategy aimed at increasing turnover and market share, but the benefits of these systems for small businesses have often been viewed with skepticism. With more limited financial and human resources, the successful acquisition and implementation of systems of this scale can be problematic for small companies; yet the potential benefits are nevertheless attractive to companies that are often continually striving to increase market share and profitability, whilst retaining a firm control on revenue costs and capital expenditure. This chapter examines two case studies where the introduction of a new CRM system was the main component of a company-wide systems upgrade or replacement. The two case studies examine the issues that have produced different degrees of success at these companies, which not only implemented new systems but also addressed the required upskilling of staff and re-engineering of core business processes.

INTRODUCTION

Construction is currently one of the largest sectors in the UK economy, with an annual turnover of £370 billion and employing 3.1 million people - 9% of the total UK workforce (Department for Business, Energy and Industrial Strategy, 2017, para. 2). The industry has long been a focus for innovation in the deployment of new materials and construction methods, and the application of innovative designs and architectural concepts. In recent years, the advent of the internet and the evolution of new routes to market have been the catalyst for further innovation for the main protagonists in the construction and property management supply chain. The UK Government has recently agreed a "Sector Deal" with the industry that "will substantially boost the

DOI: 10.4018/978-1-5225-7408-8.ch002

sector's productivity, through greater investment in innovation and skills" and "reduce the environmental impact, improve the efficiency and reduce the whole life cost of new projects" (Department for Business, Energy and Industrial Strategy, 2017, para. 3). This chapter examines two case studies of how technology transfer projects introduced new systems and innovative practices into two small-to-medium sized enterprises (SMEs) operating in this sector, as they tried to adapt to rapidly evolving business and technology environments and changing customer needs.

The first case study is at Pegasus Retirement Homes Ltd located in Cheltenham, where the company's business model centered on the outsourcing of many of its core processes, with the senior management team remaining as the main employees of the company. The second case study is of Building Solutions Ltd, a family owned house builder based in Gloucester. In both companies, the implementation of a new CRM system was a key element of the overall technology transfer project. The CRM systems concept pre-dates the widespread adoption of the internet by companies and organizations. It was originally seen as an integrated sales, marketing and service strategy, but this activity has taken on an increased importance in many organizations with the increasing use of the internet (Chen, 2005).

NEW CUSTOMER RELATIONSHIP MANAGEMENT SYSTEM AT PEGASUS RETIREMENT HOMES, CHELTENHAM

Background

Pegasus Retirement Homes plan and build private houses and apartments for sale to retired people throughout the UK from a central base in Cheltenham. In 2002, as part of their expansion programme, they wished to retain a central office in Cheltenham and use staff working from home to manage their regional operations. This would require a significant improvement in their communication systems and project monitoring capabilities. If their business processes were not improved, their expansion plans would inevitably be adversely affected. In addition, Pegasus required slicker customer service and were seeking a new outsourcing arrangement with a new fulfilment house.

At the time the company embarked on the project, there were almost 5 million owner-occupier households of retirement age in the UK. In contrast, there were only 500,000 properties geared for living in retirement – otherwise termed "sheltered accommodation". Of these 500,000 properties, only 70,000 had been built by private sector companies, and these companies were then building only 2,500 new units a year. There was thus considerable scope for the private sector to help meet the shortfall in sheltered accommodation for some years.

In the private sector, the market leader provided about 75% of the 2,500 new units constructed each year. The remaining 25% was provided by a number of companies, and Pegasus was one of the biggest of these. Pegasus sold 44 properties in 2001 and 84 in 2002, 98 in 2003 and 140 in 2004, and the sales target sales figure for 2008 was 700. The aspiration was to increase market share to 10% in 2005 and 20% in 2008. Such rapid expansion would traditionally involve setting up new branch offices and decentralisation through geographical distribution, with loss of control and spiralling costs being serious threats in a period of rapid growth. The company wanted to address this issue and expand its business without exposing themselves to these threats. At the same time, the opportunity to embrace new technologies offered additional incentives to streamline the existing operations.

Project Objectives

The overall objective was "to develop and implement new management processes and support systems that will allow the planned expansion of Pegasus in volume and geographic spread" (Teaching Company Directorate, 2002, p.2). A major challenge was to control the overheads involved in the company's rapid expansion. The company wanted to remain in Cheltenham but to allow staff to work from home to manage their expanding regional operations. The project proposal noted that "the development of management systems and processes to support a new organisational structure requires careful development both in terms of procedures and information technology" (Teaching Company Directorate, 2002, p.3).

There were a number of inter-related objectives to be delivered over a 24 month period. These were noted in the Final Report as:

- "Introduce a new Customer Relationship Management (CRM) system
- Improve the use of existing IT via training and best practice.
- Extend the use of IT to the sales office at each major development site.
- Reduce the cost of connectivity by introducing Virtual Private Network (VPN) and Wireless technology" (Momenta, 2005a, p.2)

Project Implementation

The Associate acted as the company systems development manager for two years, from January 2003 till January 2005. At the outset, a series of "quick wins" were identified, one of which was to raise the profile and perception of IT by introducing a centrally managed contact database using Microsoft Outlook. The Associate worked with secretaries and administrative staff to get them to adhere to standard ways of working and data sharing. The database was then used for mailing to circa 400 contacts at a time. It helped Pegasus staff see IT as an enabling tool that could help them in their jobs and also highlighted the need to adopt standards. Business improvement workshops were conducted by the Associate, to review current processes. Immediate results were obtained with training on how best to use and share existing IT systems in different process areas, including standardised data input protocols. Top line business processes were mapped and a number of improvements were made across the two year period of the project.

However, the main focus of the project was to evaluate the options for a new CRM system. The CRM sub-project started in November 2003 and, working with the newly appointed business systems director and other company personnel, the Associate managed the requirements gathering and package selection processes (Table 1). The detailed specification document encompassed both technical (Table 2) and functional requirements. In line with Pegasus's business model, the software selection process was undertaken in liaison with the company's outsourced order fulfilment house, based in Glasgow. Software demonstrations that used realistic data and business scenarios were held, enabling Pegasus to identify key differences in approach, cost and technology support. A short-list of two potential software vendors was agreed and reference sites visited and evaluated. Finally in June 2004, a new CRM system was selected and final testing was undertaken in liaison with the software supplier and fulfilment house. Configuration changes were made in liaison with the software supplier on points of detailed design. The use of the new system went live in November 2004 and represented a step change in Pegasus' ability to analyse marketing and sales data and hence increase sales. The new software was accessed by Pegasus staff at their main office in Cheltenham and at each sales office.

The Pegasus communications network was upgraded to VPN and wireless technology, part of a total new investment of some £250K in technology and software development. Wireless technology was introduced into the sales

Table 1. Package selection process for new CRM system (Pegasus Retirement Homes, 2004, p.4)

1. Decide Pegasus Requirements High level objectives Detailed functional requirements Rate requirements as "must have" and "nice to have" Distinguish between what is needed now and in the near future	The objective is to choose both new software and a partner
2. Document the requirements	This document will be revised and become the specification given to vendors
3. Get the senior team to formally review and agree requirements	
4. Issue specification to all prospects	
5. Review tenders and consider which software should be demonstrated.	
6. Ask each prospect to confirm that their software will meet the requirements and, if so for them to demo the software in Cheltenham	Senior management to attend demonstrations.
7. Compare software and vendors	A set of criteria will be defined for software and another set for vendors.
8. Shortlist to 2 products	Use criteria and scoring system
9. Visit reference site	It is vital to see the software in action and to be able to ask questions
10. Meeting with vendor to ask final set of detailed questions	
11. Choose software and vendor	Agree implementation programme

offices and the sales managers were trained in its use by the Associate. This sub-project completed in November 2004 to coincide with the new CRM system going live. For the first time, Pegasus employees could communicate via e-mail, whether they were in head office, a sales office or working from home. This increased overall efficiencies, with the Associate providing additional training for the sales managers in the use of personal productivity tools (email, word-processing, spreadsheets and graphics) and newly upgraded laptop hardware.

Project Outcomes

As a result of this investment in a new CRM system and associated infrastructure, Pegasus required fewer staff to manage and support an expanding sales team, reducing overheads by £150K per year. The completed project increased the rate of sale through enhanced CRM capability and improved communication to staff and customers. The new CRM system significantly improved management information that was used to drive expansion of the company's customer base. The new IT communications infrastructure reduced

Table 2. General and technical requirements specification for the new CRM system (Pegasus Retirement Homes, 2004, p. 7). 'M' indicates mandatory, 'I' indicates important and 'D' indicates desirable

Requirement/feature	Rating	Response	Comments
All of the modules of the system should be tightly integrated with one another.	M		
The system should be Internet enabled and suitable for use over a basic ADSL line (client end)	M		
Ability of the system to link to an Intranet and facilitate irregular users sending information to the system and obtaining reports and other information from the system.	M		
The system should be able to link directly to MS Office, for mail merging prospect names and addresses and other data into Word documents.	M		
Ability to generate reports that are clear, comprehensive and accurate.	M		
User defined standard reports available on line.	M		
Ability to write reports which can interrogate data from across modules.	M		
Ability to have user-customised reports that can be maintained separately.	M		
The system should be a true Microsoft Windows designed 32-bit system and not simply old code with a new Windows 'look and feel' front end.	M		
The system should use an industry standard database at the server, with SQL Server as the preferred options.	M		
The data capacity should be scalable for the number of prospect details to increase from 20,000 to 300,000 over 5 years	M		
An archive facility should be available for removing older details	I		
The system should also be capable of running under Windows XP (client) and the database capable of running on a Windows 2000 based server	M		
The system must be capable of performing selective, incremental and full back up and recovery.	M		
The supplier should specify whether they are able to provide remote access to facilitate their support of the system. Pegasus need to know exactly how remote support will be provided.	I		
The system should have workflow facilities as an option for future installation.	D		
Ability to 'drill down' through an overview screen into the underlying individual details.	D		
The system should have full support for multi-user and multi-site operation.	M		
For speed of operation when entering telephone enquiries, it should be possible to enter all details by the use of simple keyboard commands, with no use of the mouse.	D		
The system should have a context sensitive on-line help facility.	D		
All modules of the system should support searches by key data, e.g. name, post code & 'fuzzy' matches (wild card).	M		
The system should provide sufficient security of access to allow user accounts to be defined for each type of user and segregation of duties to be maintained.	M		
Ability to restrict access by organisational limits (e.g. site).	M		
Ability to provide a system log for the systems administrator.	I		
It should be possible for the system administrator to selectively update the system log without having to delete the entire log e.g. delete all log entries over a month old.	I		
Ability to record details of every letter posted.	M		
Ability to restrict access to system set up options	M		
Ability to concurrently process multiple routines/reports.	M		

overall connectivity costs, with the VPN and wireless technology being installed in every new sales office as new property development sites were set up. This was a major benefit for the fifty percent of the company's employees who worked away from the Cheltenham office, either at a development site or working from home. The project contributed to the company achieving corporate expansion on a national scale: annual turnover grew from £12.2m in 2002 to £40m for the financial year ending March 2006, achieved without any significant increase in staff overheads.

The completion of this project can be seen in the context of the enormous opportunity offered by the projected expansion in the UK retirement housing market, which was estimated to double between 2001 and 2040. The company business systems director noted at the time that "as a result of the new CRM system, it is reasonable to expect an increase in the rate of sale equal to approximately one unit per site per annum. For ten sites, this would yield an extra £800K gross profit per annum. As the number of developments for sale increases, then this figure will increase accordingly" (Monk & Wynn, 2006, p.1).

Analysis

The project was given a B grade by Momenta, the UK Government funding body, which reflected the generally successful outcomes noted above. A key turning point in the project came after 6 months with the appointment by the company of a business systems director to work alongside the Associate. This gave the project renewed impetus and direction, and contributed significantly to the requirements specification and software selection process for the new CRM system. The combined skills and experience of these two IT professionals underpinned a successful outcome to a complex technology and business project.

The Associate was involved in a range of significant technology initiatives across the two year period that required a high intensity of knowledge transfer and strong project management. Her Final Report (Momenta, 2005b) recorded that "the project benefitted from the tight control that was a feature of this company's management style", and that "there was strong project management and ownership, and the project was fully aligned to the company's business imperatives at the time. It was the catalyst for a major change in the way in which sales operations were undertaken" (p.2).

Figure 1. Change factors in the Pegasus CRM project

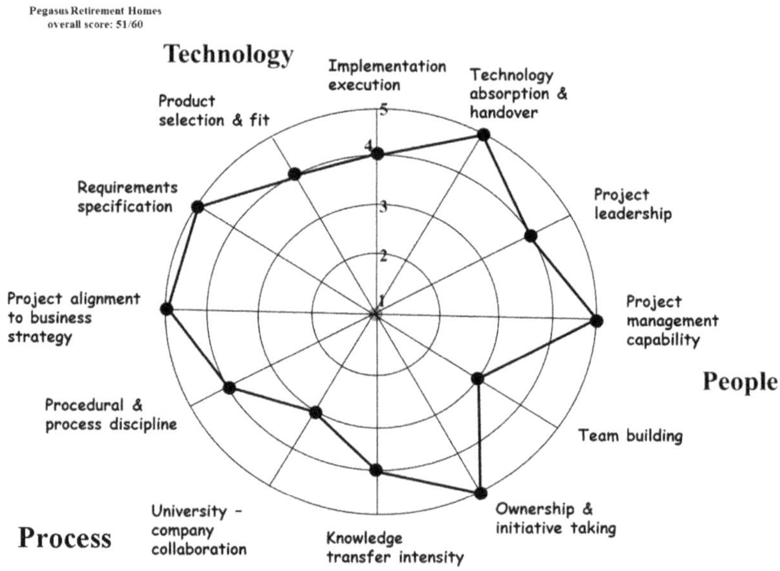

This is reflected in the change factors diagram (Figure 1). The project was viewed as critical to the delivery of the company's business vision and 5-year plan, and the importance of the project was underscored by the managing director chairing the quarterly Project Board meetings. As the Associate noted above, a strong project management ethic and issue ownership typified the way the company operated, and this was clearly evidenced in the technology transfer project. In fact, the full ownership and direction of the project by the business systems director to some extent diluted the degree of university-company collaboration, but did underpin a very successful implementation execution with strong procedural and process discipline.

"BEST OF BREED" SYSTEMS IMPLEMENTATION AT BUILDING SOLUTIONS, GLOUCESTER

Company Profile

Building Solutions was a house builder at the upper end of the construction market in quality terms, but had a relatively small market share when the project commenced in 2005. Based in south Gloucester, the company then employed 65 staff, and had hitherto purposefully tried to control its growth to sustain the quality of service. After 35 years of trading in the semi-rural district of Gloucestershire, the company was generally well regarded by relevant "specifiers" i.e. entities such as local authorities and defense contractors who may recommend or select specific construction companies. The company's recently attained Investors in People qualification and ISO 9002 accreditation had helped enhance the company's standing.

The company's business plan targeted a growth in turnover from £5.8m in 2005 to £10m by 2010. Senior management considered that there were significant opportunities for growth, but that they were being held back by outdated and ineffective use of information technology. In past years, the company had been ahead of its competitors in the deployment of IT, running a fully networked system for over twenty years; but recent growth had outpaced the development of both the IT systems and the improvement of outmoded processes in the company.

Project Rationale

Ineffective business processes had become a barrier to growth, exacerbated by standalone information systems and multiple versions of key data. Resultant problems included a lack of shared estimating intelligence, fragmented project planning, inconsistent materials purchasing, inaccurate job costing, inadequate document management and record keeping, weak customer relationship management and outdated, paper-based internal knowledge sharing. The Associate summarised the problems in his Final Report as follows:

- *"Data was being re-keyed into three disparate information systems.*
- *Poor systems management had led to one database being replicated twice, with some staff members updating one database and the rest updating the other (which was causing all sorts of confusion).*
- *Poor data management had led to inconsistent, incomplete, erroneous and superfluous information being entered into the system.*
- *System constraints had led to many manual paper based workarounds which where temporary in nature, time consuming and ineffective.*
- *All management analysis was being conducted manually (one survey conversion report took 3 admin girls 1 month to prepare).*
- *All information about one job was stored in one folder – allowing only one member of staff to access information about the job at any one time; this caused particular problems because job folders were frequently*

> *taken off site, left under peoples desks and their whereabouts to other staff members unknown" (Momenta, 2007a, p.3).*

One of the company's major sub-sectors was insurance reconstruction. These clients had highly developed systems and expected timely, accurate information on the status of contracts, together with rapid processing of instructions and invoices. The company's systems were at capacity and there was an urgent need to redesign the relevant processes and implement reliable, expandable new systems. Whilst the company Board appreciated the value of completely reviewing core processes, it recognized that the necessary skills did not exist within the company to deliver a long-lasting sustainable solution that would be embedded in the business. Adopting well-conceived improvements to these processes, together with the correct implementation of new technologies and training, would enable the company to achieve its growth and profit targets and improve customer service, with a resultant improvement in staff satisfaction. The overarching objective of the project was thus to "install new information systems and technologies and reengineer business processes to drive through 5% cost reductions and a 15% increase in company turnover per annum" (Momenta, 2005c, p.2). Given that the company already used the Sage financials package, it was decided at an early stage to adopt a "best of breed" strategy through which new packages would be selected for each major business function. Issues of integration were to be addressed on a package by package basis.

Project Implementation

Based on the outline project plan (Figure 2) contained in the project proposal, the Associate led a two stage implementation spanning 22 months, completing in March 2007. Stage 1 established a stable technology platform for the company's current and future information systems. This was an urgent priority as investment had been restricted for several years. Desktop personal computers were of variable standards, there were multiple versions of desktop software and there was no clear product policy. The local area network bandwidth was inadequate even for systems running at the beginning of the project.

The Associate drew up a technology upgrade plan, gained budgetary approval for new desktop hardware and software and a major network upgrade. This was completed within two months of the start of the project, and involved significant "out-of-hours" working.

Figure 2. Initial project plan for the technology transfer project at Building Solutions (Momenta, 2005c, p.38)

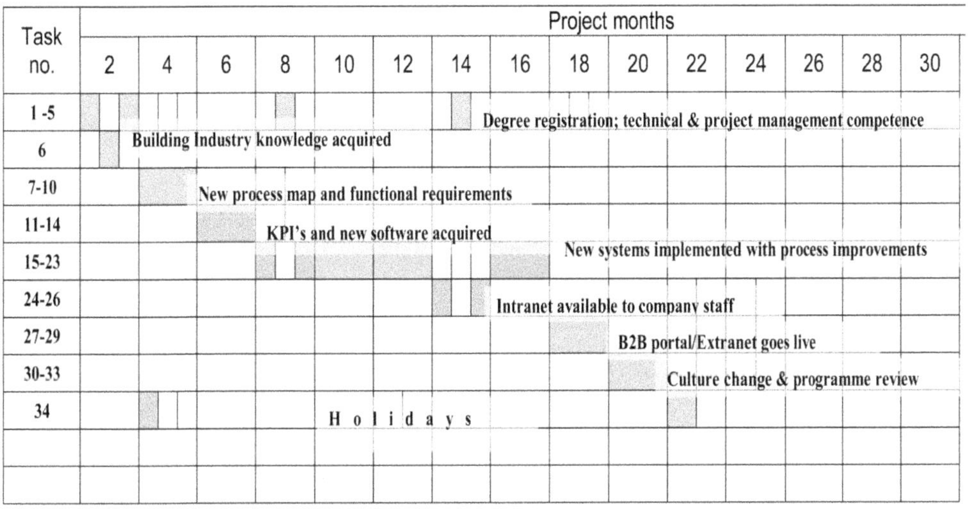

Stage 2 took a longer-term view of the systems required for the company to support its growth over the next five to ten years. The Associate familiarized himself with the company structure and got to know the significant number of field based workers, by running business improvement workshops and working with senior management to see how technology could support business change. This investigation identified areas for improvement and underlying technologies that could be used to facilitate new systems. These were developed into an information systems strategy that was presented to the Board in June 2006. Following Board approval, the Associate managed the procurement of new software packages in line with Board directives. The implementation roll-out had several main phases with investment across the two project stages in technology infrastructure (networks, servers, desktop PCs) and new software exceeding £60K. Senior management elected to build on the existing investment in current systems that were deemed strategically sound (Sage and Commander) and to add point package solutions to provide an overall integrated "best of breed" systems portfolio. Stage 2 of the project comprised the following main software and infrastructure components (Figure 3):

Figure 3. New and existing systems at Building Solutions, 2007

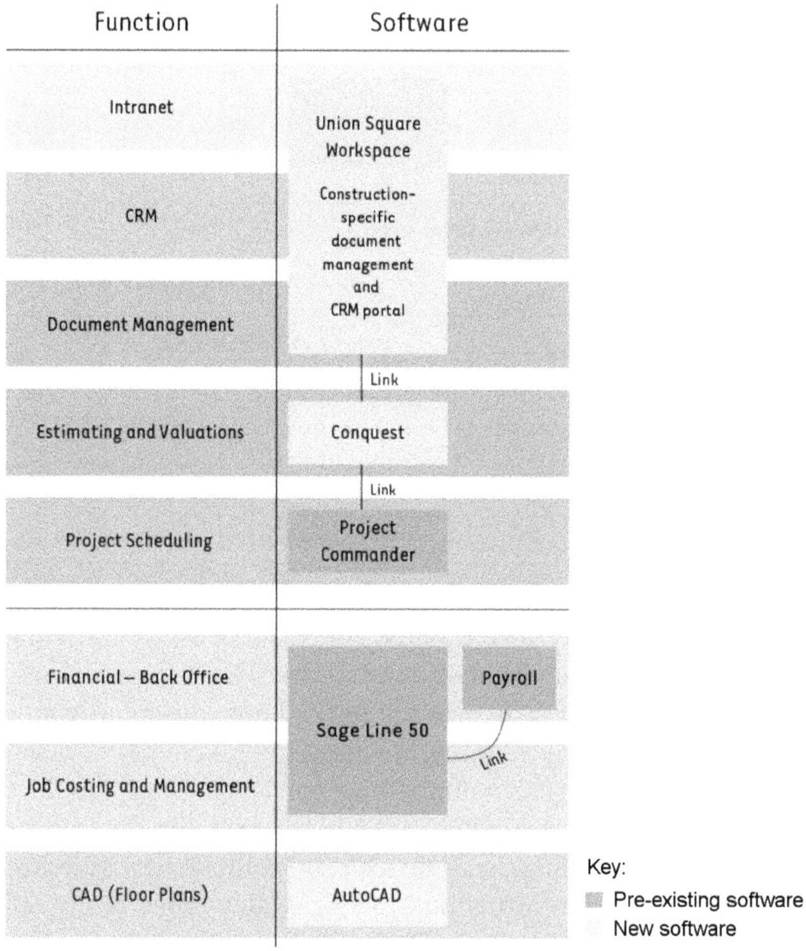

- Implementation of a new CRM software package (Union Square), replacing the existing system which was not widely used within the company and which had limited functionality.
- Implementation of a new project estimating system (Conquest).
- Deployment of a new package (AutoCAD) for detailed floor planning and design.
- Improvement of the use of desktop personal productivity software via training and adoption of best practice standards.
- Extension of remote communications via a virtual private network (VPN) to allow connection to key business partners, sales staff and field based managers.

In the early days of stage 2, the Associate identified a series of "quick wins" that helped raise the profile and perception of IT by cleansing the data in the two legacy customer databases. This laid the foundation for the introduction of the new CRM system, and the new estimating software in the latter half of the project. The Associate managed the requirements gathering and software selection process, involving software demonstrations from third party vendors that used realistic data and business scenarios, thus enabling the company to identify key differences in approach between vendors. This was critical in eliminating one of the other mainstream packages that had been a front-runner in the early stages.

The resultant CRM implementation ran in parallel with the estimating systems project for the final 6 months of the project and represented a step change in Building Solutions' ability to analyze marketing and sales data and increase sales. The Associate took responsibility for testing the new systems and liaising with the software suppliers regarding systems configuration, bug fixing, change requests and supplier consultancy. The Associate also organized and led training in two main areas:

- Training to consolidate the changes made in internal processes, and to ensure that best use was made of the new systems.
- Training and support of the key users of the new CRM system, to embed the system within the company and allow on-going support by company staff. The Associate took an innovative approach here, and authored a range of video based training materials that were made available on personal computers to all users to simulate key functions on the new system.

Project Outcomes

At an operational level, the company developed the capabilities to use and onward develop a new range of applications to support the company's sales processes and provide critical management information, notably in the cost estimating of projects. Building Solutions had upgraded its IT infrastructure to improve systems performance and allow remote field working from its new sales office in Pittville in Cheltenham and connectivity with sales staff working in show houses in the region. This allowed the company to grow its turnover without significant increase in staff overheads. Annual savings were estimated at £15K.

The company also benefited from the new capabilities of the CRM system and improved networks, increasing the rate of sales, estimated at constituting one new major contract per annum and yielding an extra £750K turnover a year (a net profit of circa £37.5K). Further benefits came from the development of project estimating capabilities, eliminating the re-keying of data and providing an on-line snapshot of project financing. This gave the project contract managers a quick and effective feedback tool to guide their pricing and eliminate painstaking manual work in working out contract detail. Company procedures were changed, including those governing job and cost allocation, and cost verification and analysis. This "produced improved visibility of project costs and has given a better measure of project profitability" (Momenta, 2007b, p.5).

The project was also the catalyst for the introduction of a new management structure. The critical examination and analysis of existing systems led to Board level agreement to significant process change. One main feature of this was the merging of all project contract managers, operatives and support administrators into one directorate. The company's improved technology standards and capabilities meant that they could interact with a wider range of clients (for example, in the retail sector, which is very demanding on the compatibility of their supplier's IT systems). The ability to better serve the specifier and insurance markets, through greater electronic supply chain integration, also enhanced the job satisfaction of all concerned. The managing director of Building Solutions stressed the importance of the project "in providing a key platform for our growth aspirations over the next five years" (Monk & Wynn, 2008, p.1).

Analysis

The project was given a C grade by Momenta, the UK Government funding body, in line with the generally successful outcomes noted above. The technology elements of the project were sound, but there were some issues regarding team building and people skills. The Associate had to adjust to the building industry culture that required strict procedural and process discipline, and university-company collaboration was acceptable but could have involved knowledge exchange on a broader range of topics. Nevertheless, the company provided strong leadership and ensured the handover was completed and that on-going support for the new technology elements was in place. Overall, the project achieved its objectives, and provided a sound systems and technology platform for the company's future growth plans.

The change factors diagram (Figure 4) highlights the significance of project alignment with business strategy and strong project leadership, particularly from within the company, as the two main factors that underpinned project success. These factors more than made up for some weaknesses in procedural discipline and team building in particular. Overall, the requirements specification and product selection for the new packages were well done and implementation executed effectively. Ownership and initiative taking were in evidence, notably in the Associate's leadership of well-planned and effectively-conducted training programmes with company staff. Technology absorption and handover was evident in many parts of the business. The Final Report (Momenta, 2007b) noted that "many aspects of day-to-day operations have been affected" and that "the ability of Building Solutions to respond to customers' information and collaboration requirements has numerous benefits in maintaining existing relationships and securing new business. This has reduced the overhead on management in resolving data conflicts in the customer area and generally improved information quality and speeded up effective decision-making" (p.4).

Figure 4. Change factors for the Building Solutions systems project

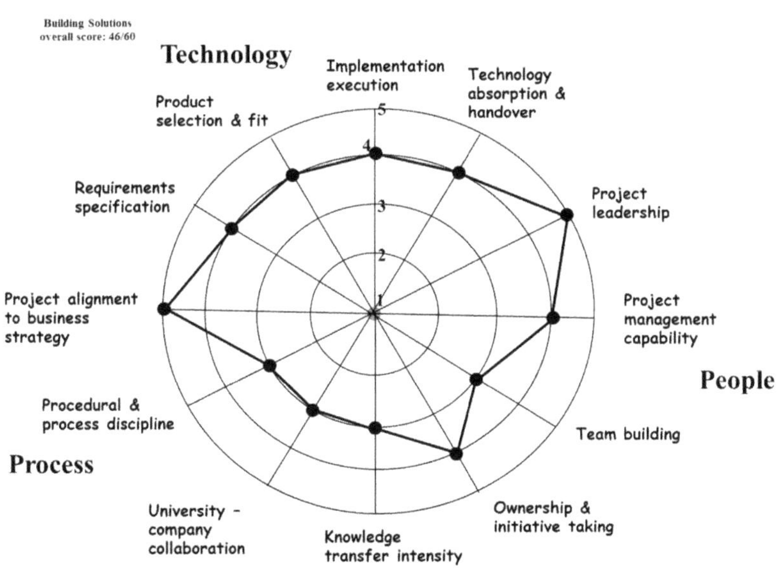

CONCLUSION

Since the turn of the century, CRM systems have emerged as a major component of information systems strategy as companies have realized the importance of integrating all customer facing systems as a component of their business strategy, to manage and build stronger customer relations. Initially, CRM adoption was largely restricted to the bigger organizations that had the necessary resources to identify the opportunities, define requirements and implement new processes and technologies that were often part of a broader e-business strategy. More recently, however, an increasing number of SMEs have embarked upon CRM projects, attracted by the possibilities of growth and innovation in their customer related activities and processes.

These projects are good examples of how this manifested itself in the construction industry. They both focused on introducing new technologies to improve internal processes, and in particular, customer facing operations, using new CRM systems (Figure 5). These two house-builders saw the opportunities for implementing CRM capability and the potential for process change to support bottom-line increases in turnover and profit. The Pegasus project was somewhat more radical and included a major change in networking technology as well as a new outsourced CRM capability. It supported a dramatic increase in turnover in the years following the project. The Building Solutions project was more of an incremental, step-by-step change in systems capability, but made a significant contribution to process improvement and infrastructure upgrade in the company.

Figure 5. House builder projects: technology transfer focus and degree of innovation

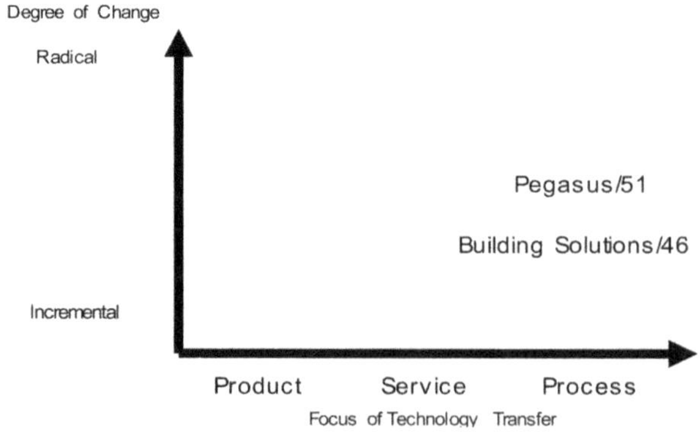

REFERENCES

Chen, S. (2005). *Strategic Management of e-Business* (2nd ed.). Hoboken, NJ: J Wiley.

Department for Business, Energy & Industrial Strategy. (2017). *Construction Sector Deal.* Retrieved September 12th, 2018 from: https://www.gov.uk/government/publications/industrial-strategy-sector-deals/introduction-to-sector-deals

Momenta. (2005a). *Knowledge Transfer Partnership Final Report KTP 4113 (Pegasus Retirement Homes).* TTI Database Record, KTP Programme Office. (Unpublished)

Momenta. (2005b). *Knowledge Transfer Partnership Associate Final Report KTP 4113 (Pegasus Retirement Homes).* TTI Database Record, KTP Programme Office. (Unpublished)

Momenta. (2005c). *Knowledge Transfer Partnership Project Proposal and Application Form Building Solutions.* TTI Database Record, KTP Programme Office. (Unpublished)

Momenta. (2007a). *Knowledge Transfer Partnership Associate Final Report KTP 0807 (Building Solutions).* TTI Database Record, KTP Programme Office. (Unpublished)

Momenta. (2007b). *Knowledge Transfer Partnership Final Report KTP 0807 (Building Solutions).* TTI Database Record, KTP Programme Office. (Unpublished)

Monk, C., & Wynn, M. (2006). *Knowledge Transfer Partnership Case Study 1 - Pegasus Retirement Homes.* DTI/Momenta/University of Gloucestershire. Retrieved September 13th, 2018 from http://eprints.glos.ac.uk/5528/

Monk, C., & Wynn, M. (2008). *Knowledge Transfer Partnership Case Study 4 - Building Solutions.* Technology Strategy Board/University of Gloucestershire. Retrieved September 13th, 2018 from http://eprints.glos.ac.uk/5531/

Pegasus Retirement Homes. (2004). *Prospective Purchaser Tracking System Specification (Project Klondike).* Pegasus Retirement Homers. (Unpublished)

Teaching Company Directorate. (2002). *Teaching Company Scheme Grant Application and Proposal Form Pegasus Retirement Homes.* Teaching Company Directorate, Brunel House. (Unpublished)

KEY TERMS AND DEFINITIONS

Best of Breed (BOB): An information systems strategy that involves finding the most appropriate software system for each function in a company or organization. It is an alternative to acquiring an Enterprise Resource Planning system that supports all (or most) business processes.

Customer Relationship Management (CRM): The combination of sales and marketing activities that focus on fulfilling customer requirements. There are many standalone CRM software systems, but they may also be part of an integrated Enterprise Resource Planning system.

Investors in People (IIP): An award to a company or organization that recognizes high standards in the management and development of its people. From 1991 to 2017, Investors in People was owned by the UK Government but was transferred in February 2017 to the Investors in People Community Interest Company.

ISO 9002: A standard developed and published by the International Organization for Standardization. It is a non-industry specific certification for quality assurance in production, installation, and servicing.

Virtual Private Network (VPN): A technology that allows a company's remote users and branch offices to securely access their corporate systems, providing a safe and encrypted connection over a less secure network, such as the internet. It enables users to send and receive data across public networks as if their computers or hand-held devices were directly connected to the private network.

Chapter 3
Small Software Companies

ABSTRACT

The technology industry is dominated by major international companies, but there are also many SMEs, SBEs, and micro-companies operating effectively in this industry sector. Maintaining innovation is a key challenge, especially for the small players in this market, and technology transfer to develop new products and services is particularly challenging. This chapter examines four technology transfer projects in small software companies, three of which focused on new product development, and the fourth on a new service provision. All four projects were generally well managed, but effecting this degree of change requires more than good management, especially in such small companies. It is not surprising that some of these projects failed to achieve their objectives in the mid-term.

INTRODUCTION

Infiniti Research (2018) recently noted that technology companies "need to keep themselves updated and incorporate new technologies to survive in this highly competitive market where the number of big and small players are increasing continuously" (para.7). Another major challenge is to keep up with changing customer demands. "Players in this industry are continuously keeping themselves updated with the changing demands of customers and are focusing their efforts towards developing new solutions to meet these requirements" (para.1).

This chapter examines four technology transfer projects in small software companies that aimed at meeting one or other of these objectives. The first

DOI: 10.4018/978-1-5225-7408-8.ch003

case study at E-Business Services, a small software house in Hereford, uses aliases for the company itself and the two major customers that feature in the project (on the request of the company managing director). The other cases are at AuraQ in Malvern, which at the time of the project was a micro company providing software services, and at Muddy Boots Software, a well-established SME with successful software products sold into companies operating in the food industry supply chain. The technology change project at Beaumont Travel is also included in this chapter, because it became the catalyst for the company's transition from a coach and bus company to a software provider.

NEW WEB-BASED PRODUCTS AT E-BUSINESS SERVICES, HEREFORD

Company Profile

E-Business Services (EBS) was a micro company of less than 10 staff, based in Hereford, Herefordshire, when it embarked on the technology transfer project in 2006. At that time, this small software house was generating revenue from the delivery of bespoke and packaged web based products. The bulk of company turnover came from the sale of its main web product (Woolpack), which was used by client companies as an Intranet, information portal and communications management tool with their end customers. The technology transfer project focused on the development of new technology products, using web-based software tools.

Project Objectives

In 2006, EBS initiated a new business plan to move away from purely bespoke adaptations of their existent information portal product (Woolpack), towards a company specialising in custom built web portals for key customers running Microsoft Dynamics (previously called Navision) as their core back-bone system, especially in the financial and related service sectors. The project "was key to implementing this business transition as it delivered the first attempts at design and development of web portals in this niche market" (Technology Strategy Board, 2008a, p.2). This project aimed to use new technology developments to innovate the company's product offering.

EBS had worked for a number of years for Skindal Life, a financial services company with many branches around the UK. They were viewing the possibility of sharing data with other parties as they entered into new partnerships with a number of other well-known financial services institutions. These partners sold financial products administered by Skindal Life through their established adviser channels; however the partners wanted their advisers to access these plans and products through their web site. In addition, to attract potential new investors, the solution needed to have the capability for "fund supermarket" trading.

New Product Development: Project 1

The Associate joined EBS in late 2006 and after initial familiarisation with the company's existing products and clients, he embarked on the requirements analysis for the new Skindal Life portal product. Working with the company development team, a portal architecture diagram was developed, and source code was written and debugged. After extensive testing, the new Skindal Life portal was implemented in April 2007, involving new technologies (.net and Biztalk) and a highly secure systems architecture. Users from partner sites were directed to Skindal Life's website, using the partner's authentication, so users did not have to re-login. Users were able to transparently access pension data held in the Skindal Life back office systems. As they accessed the web pages they were dynamically branded. The net effect was that the partners' investment was minimised, but they gained access to the full product functionality. The Final Report noted that "once access to the site is provided, Partners and Advisers can view all their plans ....advisers can additionally buy and sell unit trusts online, through a real time web service connection into a fund supermarket. Data is exchanged between the website and their back office systems (Microsoft Dynamics). All exchanges are performed securely via web services using XML" (Technology Strategy Board, 2008a, p.1).

Strategically, Skindal Life was able to open new distribution channels, offering its services through partners. Additional business was achieved without an exponential growth in headcount. Operationally, Skindal Life provided access to their plan data through these channels in a way which simply could not have been achieved without the use of the internet; and the business partners shared the benefit of Skindal Life's web services, without having to replicate the functionality themselves. Advisers had the additional benefit of being able to access information 24/7. Skindal Life witnessed increasing volumes of web-based enquiries, as well as greater use of the online fund supermarket dealing service. The company were able to handle significantly increased volumes of new business, whilst continuing to provide their award-winning high quality service.

New Product Development: Project 2

Building on the experience at Skindal Life, the new web portal and middleware products were further deployed at a second customer – Effective People Management (EPM). EPM provide personnel services, consultancy and training to a large number of Primary, Secondary, Special, Foundation, Aided and Community schools and academies. EPM had extensive expertise, experience and understanding of personnel issues in schools who they helped to achieve the high level of staff management necessary for running a successful school. EPM provided these services to hundreds of schools across the UK, ensuring tens of thousands of education sector employees were paid timely and accurately every month. EPM concluded that these working practices could be greatly improved with the use of the internet, and, using EBS's new products, a new e-business strategy was developed to facilitate access to EPM's services over the internet via a secure portal enabling the schools to maintain and view their information in EPM's back office systems.

In May, 2007, a requirements document and functional specification were put together and discussed and signed off by EBS management. There were a series of phases in the development of the product (Table 1), followed by alpha and beta releases in December 2007, with the senior in-house developer performing a quality assurance role. Full testing was undertaken

Table 1. Development phases for the new web portal at EPM (Technology Strategy Board, 2008b, p.2)

Development Phase	Activity/Output
Infrastructure Phase	Research and install Message Broker service for SQL Server 2005
Authentication Phase	Develop CSS Structure of EPM Portal, Master pages
Home Page Phase	Home Page Menu design
List Phase	List all employees
View Phase	View employees
Create/Edit Phase	Create employee, edit employee
My details List Phase	My details page

in early 2008, culminating in the implementation of the new web portal at EPM later that year.

The portal (Figure 1) was a secure, confidential web site only accessible by the head teacher and authorised staff in each school. To use the site, schools were required to follow a quick and simple registration process to ensure that only authorised users had access. The web portal was developed using .net technologies and installed using EBS's Message Broker service for SQL Server 2005. Key features of the new web portal included the ability to submit, view and amend details of new employees and their contracts, submit overtime for staff and expenses for payroll, confirm absence due to sickness and holiday and view employees' payslips and P60 forms. The schools also benefited from having greater visibility of, and 24x7 access to, their school specific information, including statutory required information.

Figure 1. Architecture of the web portal built for Effective People Management

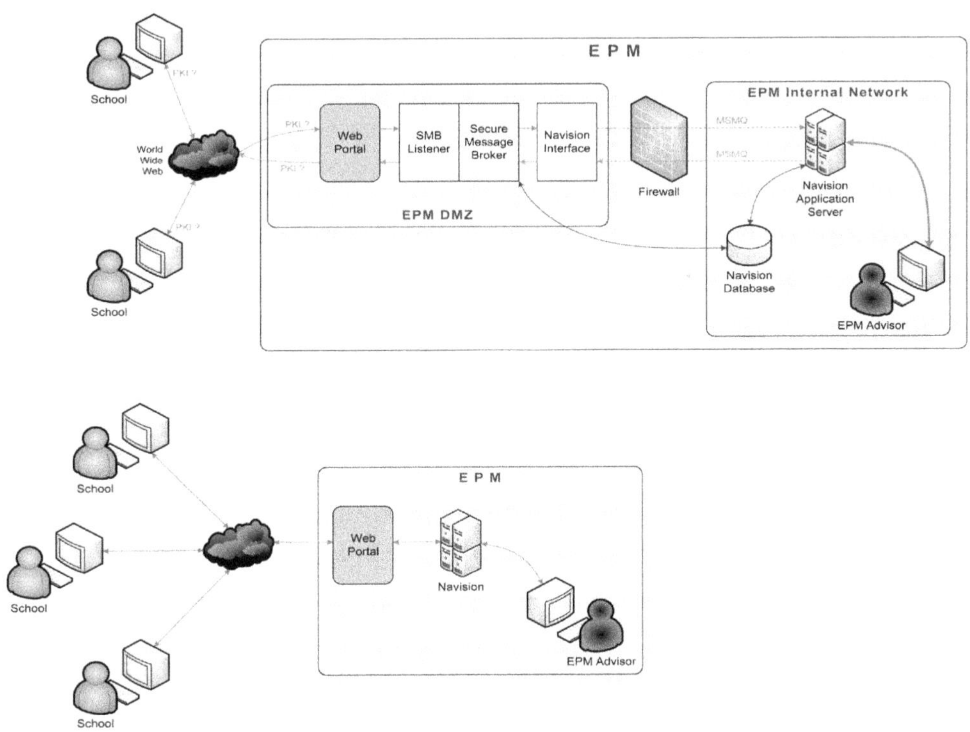

Project Benefits

EBS established "a foothold in the web portal market linked to Microsoft Dynamics" and it was "expected that this type of development will provide a major income stream in the coming years" (Technology Strategy Board, 2008a, p.3). This is reflected in the company's growth in revenue from £120K in 2004/5 to £295K in 2006/7. Key to the success of the developments at Skindal Life and EPM was an early understanding of the potential role of "middleware" software in client companies' information systems architecture. An important issue for many service companies is that their core corporate data is held in internal software packages, and is often not readily available to business partners via their websites. Companies in this sector thus tended to have problems with systems integration and often resorted to third party "middleware" products to link systems together. These products have since emerged as being very significant in overall systems strategy. They generally have no direct user interface for the end-user, but act as a "funnel" for the transport of information between applications, and provide a key link mechanism between the web-based portal and in-house information sources. The new web portal products used leading mainstream technologies - ASP.net, VB.net and SQL Server 2005. Technology transfer was multi-faceted and at various levels, supporting the company's new strategic direction, and providing the Associate, who gained his Microsoft Technical Certified Specialist (MCTS) qualification in the course of the project, with a new career path.

Analysis

Despite the successful impact of the project at EBS, it scored only moderately overall – 43 out of 60 – on the change factors diagram (Figure 2), in part because of a poor score for team building, where some of the project team members found it difficult to work well together. This aligns with the funding body's grading of the projects as C. On the positive side, project leadership from senior company personnel was a key factor - the managing director chaired the weekly project management meetings as well as the three monthly Project Board meetings with government sponsors. An effective selection and fit of new technologies to accord with requirements was also an important factor; and the application of new web-based tools and programming languages supported the development and delivery of new solutions which allowed key

Figure 2. Change factors in the E-Business Services project

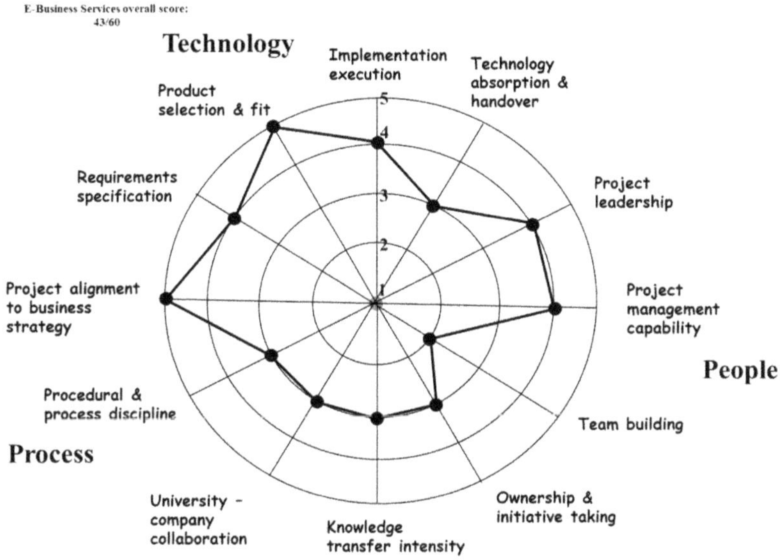

customers to make a step change in their use of the internet for customer facing processes. Execution of the project plan was done efficiently and effectively to time and budget, using appropriate project management methods; and the handover of project and technology knowledge was of significance in allowing the continued successful use of new technologies. As regards the company's future, the Final Report (Technology Strategy Board, 2008a) took the view that "the overall upskilling of staff to use .net and other new technologies" was seen to "help its competitive position" (p.6). This was done within the project duration through the structured programme of regular project meetings and related workshops.

The new technologies provided the opportunity for a change in company strategy regarding product development. Despite the team building issues noted above, university-company collaboration was at the core of the project, which was based on the two entities working together for mutually beneficial ends. This was aided by the fact that the company managers and the university supervisor had worked together in a private sector company some years prior to the start of this project. The intensity of technology transfer was of significance as company staff had knowledge of, but little experience with, the web-based tools deployed in the project. The inter-personnel issues alluded to above had some negative impacts, but this was generally well managed.

It is a testament to the technical ability and perseverance of the company management and staff that the project was finally successful in bringing new products to market that used the leading web technologies of the day.

NEW TEMPLATE DEVELOPMENTS AT AuraQ, MALVERN

Company Profile

AuraQ was established in the late 1990s (albeit under another name – NTL), specializing in process improvement. Located in Malvern, the company steadily built its turnover, generating revenues of over £0.5m in 2007 prior to the start of the project, when it had just 6 full-time staff. In 2007, the company took the decision to move towards providing software solutions rather than just process management services to its clients, and specifically to offer bespoke software components based on the Metastorm Business Process Management (BPM) product range.

Project Rationale

This technology transfer project was to be the platform for a major infusion of knowledge regarding the Metastorm products and their development for bespoke and template applications. Metastorm had a range of tools for process redesign, organizational restructuring, systems and process integration and business intelligence reporting, and could manage both human-centric and system-centric processes. AuraQ were the only accredited Metastorm partner in the UK, and this provided an opportunity to significantly grow the business in this new market niche. The company had the opportunity to establish itself as a specialist operator in the BPM market in the region, based on its position as a Metastorm partner. The intention was to make the Metastorm templates configurable for new and existing clients, and provide the company with a real opportunity to establish itself in this emerging field of process optimization and business change. The overall aim was to "research, develop and bring to market innovative business process modelling products to double turnover over a three-year period" (Technology Strategy Board, 2008c, p.1), with the initial project plan comprising 7 stages spanning a 21 month period (Figure 3).

Figure 3. The original plan for the AuraQ project 2009-11 (Technology Strategy Board, 2008c, pp.18-19)

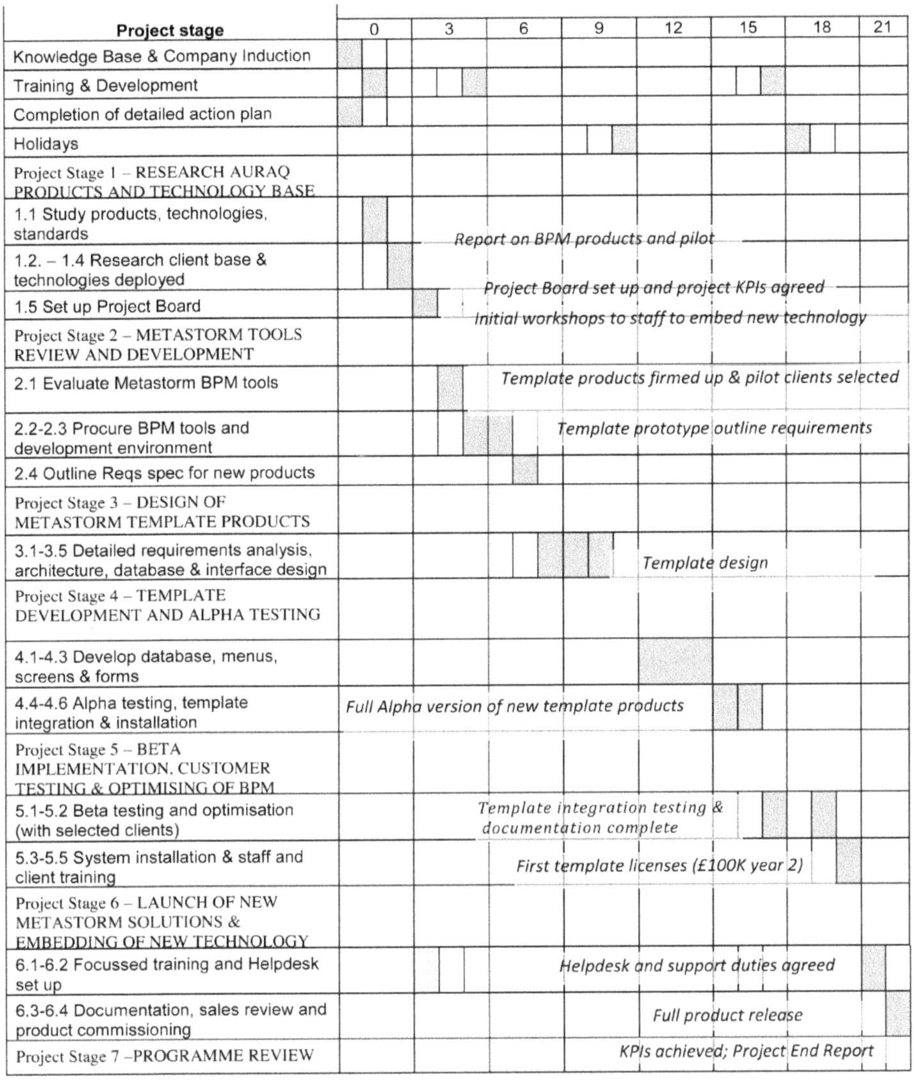

Project Implementation

The project started in 2009 and completed in 2011, and followed a standard lifecycle for such a technology project, more or less in line with the original project plan. First, the Associate researched and documented the company's business requirements to confirm what was needed in terms of Metastorm template development, and also analysed the business models of some of AuraQ's existing clients. These were then worked up into a software requirement specification, software functional specification and software design specification. Database and user interface design were then undertaken and the layered software architecture was developed. This constituted the basis for delivery of the final software templates to AuraQ, for sale to, and utilization by, Metastorm users. These software templates were aimed at being both customizable to different user requirements yet robust and scalable enough to work in different technical environments.

There were a series of delays in the project timescale, mainly because of resourcing problems within AuraQ. The Final Report (Technology Strategy Board, 2011) noted that "the project required a software resource for the development activity which was not released at the time expected. Its release was delayed for unforeseeable reasons" (p.3). However, the Associate "provided input to other areas of development which has assisted existing staff" (p.4) and by the end of the project, there had been some handover and transfer of knowledge from the Associate to company staff, but this was undermined by the non-availability of in-house staff. The knowledge areas encompassed in the handover included Metastorm development tools (Pro-Vision, Discovery and Insight), application of the .net orchestration tool to develop bespoke applications and templates, process analysis and the formulation of software requirements specifications for new template adaptation and development, and design and coding of bespoke software components in the .net environment.

Project Outcomes

By project closure, the software templates had been developed and delivered. The Final Report noted that "AuraQ can now provide a bespoke user interface for Metastorm BPM users" and that "this will be sold as a product to enhance usability and provide an interface with industry standard products such as Microsoft SharePoint". It also noted that this would "generate income from sales and support which is in addition to our usual income from services" (Technology Strategy Board, 2011, p.5).

However, these new templates were never sold by AuraQ. This was in the main because of significant changes in the Metastorm market. The Metastorm products were acquired by another software house (OpenText), who halted the expected development of the product and this severely limited the market potential of the templates developed in the project. At the same time, AuraQ had "been through a difficult trading period due to the recession" (Technology Strategy Board, 2011, p.4), with annual turnover dropping from £525K to £170K in 2010/11. Although the project had been a technical success in terms of software development, it had not been so in terms of the planned change in product offering and expected growth in turnover and profit.

Analysis

The AuraQ project aimed at template innovation with the Metastorm product, but failed to deliver against targeted objectives for a number of reasons. Although AuraQ still support Metastorm users, the envisaged expansion based on the new Metastorm templates did not materialize. Nevertheless, by forging links with new business partners, the company came through some turbulent financial times in 2010/11, and have since recovered well and now have 33 staff compared with just 6 when the KTP was started, and a turnover of £2.2m in 2016/17. This company has remained entrepreneurial, finding new development opportunities with other software providers.

The change factors diagram (Figure 4) provides some insight into where things went wrong. Although the project was initially closely aligned to the company's business strategy, by the end of the project the company was struggling to survive and the project was undermined by changes in the

Figure 4. Change factors in the AuraQ BPM project

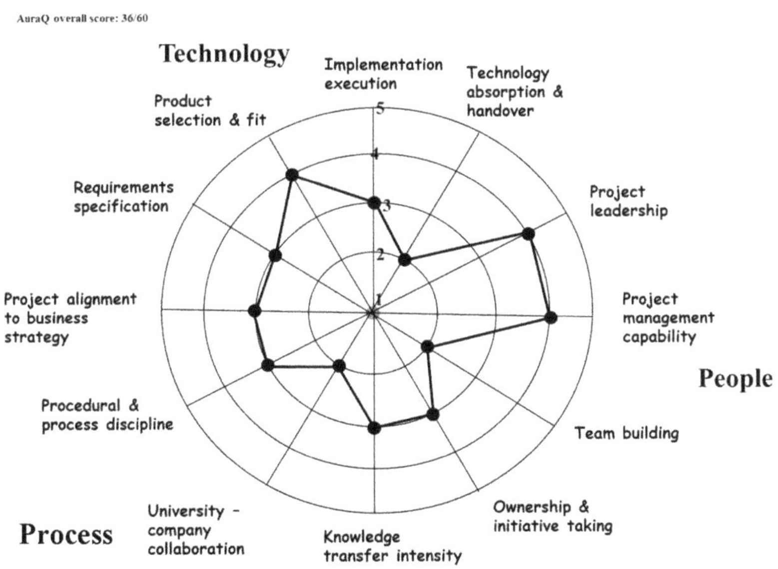

Metastorm template market. Internal resourcing was an issue across the duration of the project, and despite some strengths in project leadership and project management capability, university-company collaboration and team building were weak, with the Associate often left working alone on the project. This inevitably impacted negatively on implementation execution and technology absorption and handover. These shortcomings in the project are reflected in the D grade given to the project by the Technology Strategy Board.

NEW PRODUCT DEVELOPMENT AT BEAUMONT TRAVEL, GLOUCESTER

Background

When this project started in 2005, Beaumont Travel was not a software company, but through the course of this project, it became one. The company, as its name suggests, was a bus and coach company, located in the center of Gloucester, with 36 staff and a turnover of £1.1m. It had been built up by Mike Sadler, and his wife and son also worked in the company. In the early years of the millennium, Beaumont Travel had developed bespoke software for their own in-house use, but they had also generated some interest in their software from other operators in the industry. To further develop these products, the company realized that they needed additional technical and project management skills and thus sought support from the local university to support them in this initiative.

Project Rationale

The company management realized that, in addition to using the software themselves, there was an opportunity to sell the final software products into the transportation industry sector. The overall aim of the project was "to design, develop and market a suite of advanced software for the transportation industry encompassing central business management and a range of integrated sub-systems" (Momenta, 2005, p.2). It was noted, however, that "the current prototype products have been written by Beaumont Travel without the required knowledge and expertise in e-business system design, development and management" (Momenta, 2005, p.21). The areas of expertise that the company sought from the university included the design and development of new software using web-based technologies such as PHP, MySQL, HTML, JavaScript, and Microsoft.net and Dreamweaver, and software integration and project management skills and experience. The initial project plan for the new system (given the name "Workplace") was for a 24 month period (Figure 5).

Project Implementation

The project objectives and how they were addressed are summarized in Table 2. The initial focus was on market research (through questionnaire, interview, survey, and systems user reviews) to understand the key business requirements in the bus and coach industry and confirm the market potential of the proposed software products. This was followed by the analysis and modelling of mainstream business processes in the transportation industry, which provided the basis for the development of a web-based solution with open-architecture to allow the new software products to be used in the transportation industry.

The software was built around a new database-independent central core system (TravelManager) that was web-based and provided simple connectivity for customers operating across the internet (Figure 6). The TravelManager system included modules for central daily business management, field-bus

Figure 5. The outline project plan for the Beaumont Travel "Workplace" system project (Momenta, 2005, p.38)

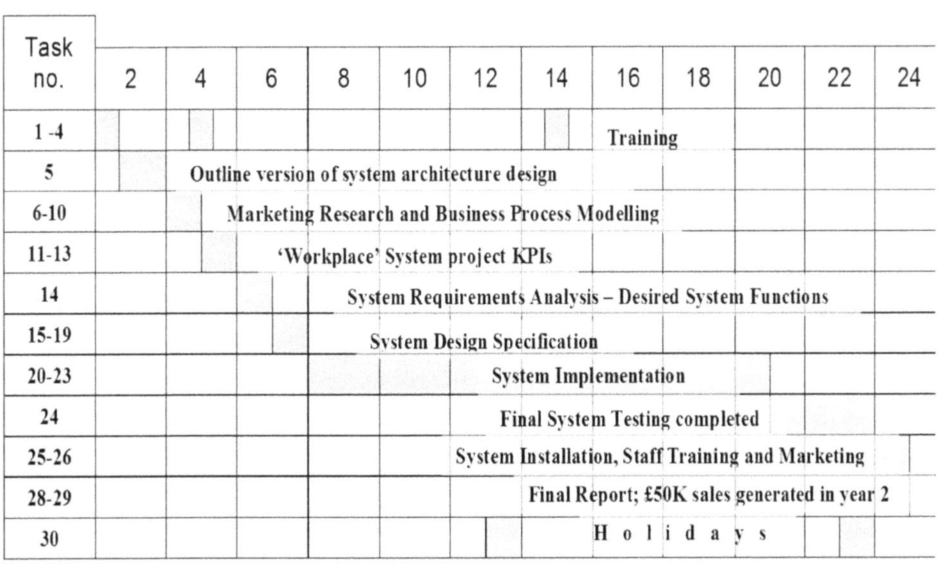

Table 2. Project objectives and outcomes at Beaumont Travel (Technology Strategy Board, 2007, p.2)

Project Objective	Outcome/Delivery against Objective
Market research through questionnaire, interview, survey, and systems user reviews to understand and identify the business requirements in the bus & coach industry.	Provided a clear picture for the Company, Associate and University of business requirements.
Analyse, model & optimise the present business processes of Beaumont Travel and other similar businesses to maximise benefits from the application of advanced e-business systems.	Provided a defined web-based solution with open-architecture to allow the applications not only to transportation industry but also to SMEs in other industries
Analyse and model key data entities.	Contributed to defining the requirements which was essential for all parties to achieve the final solution.
Propose, design and develop a new database-independent central core system (TravelManager) that will run on a wide range of servers and is operating-system independent.	The final product is web-based with simple connectivity for the customers resulting in a commercially acceptable product.
Design and develop interface modules that allow information exchange between TravelManager and vehicle maintenance systems (and other add-on modules).	All required interfaces modules have been developed to achieve system integration, and can be used by people of limited e-business skills.
Alpha and Beta testing with prospective customers, and gather feedback, build dependence and capture further requirements for future development.	The positive feedback from customers allowed us to modify some functions of the system to meet individual customer's business requirements.
Market the new software and establish the concept of Net Centric capability within the target market.	Extensive marketing activities have been carried out with huge success not only in transportation industry, but a modified working version has been selected by BT as their key CRM system (BT Bizbox) for SMEs.

Figure 6. Systems architecture for the TravelManager software product (Wynn & Zhang, 2008)

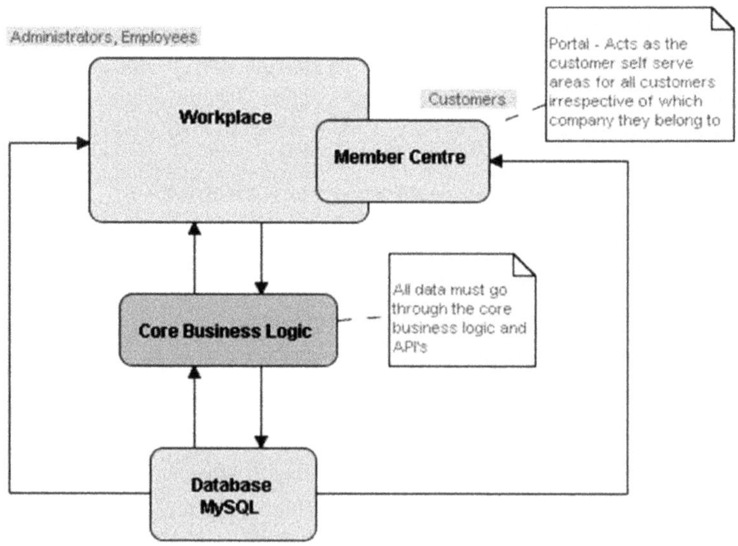

passenger and vehicle maintenance information. Interface modules were added to allow information exchange between TravelManager and other third party systems. These modules were alpha and beta tested with prospective customers leading to modification of some functions of the system to meet the individual customer's business requirements.

Project Outcomes

According to the Final Report, the company had "been successfully transformed from a traditional bus and coach service providing company to an advanced IT-centered company", and "the new knowledge gained from the project has significantly enhanced its ability to design, develop and market comprehensive software packages" (Technology Strategy Board, 2007, p.3). Feedback from early pilot customers suggested that their operational efficiency had improved from deployment of the TravelManager system (Figure 7).

Figure 7. The TravelManager software product prototype 2007 (University of Gloucestershire, 2007)

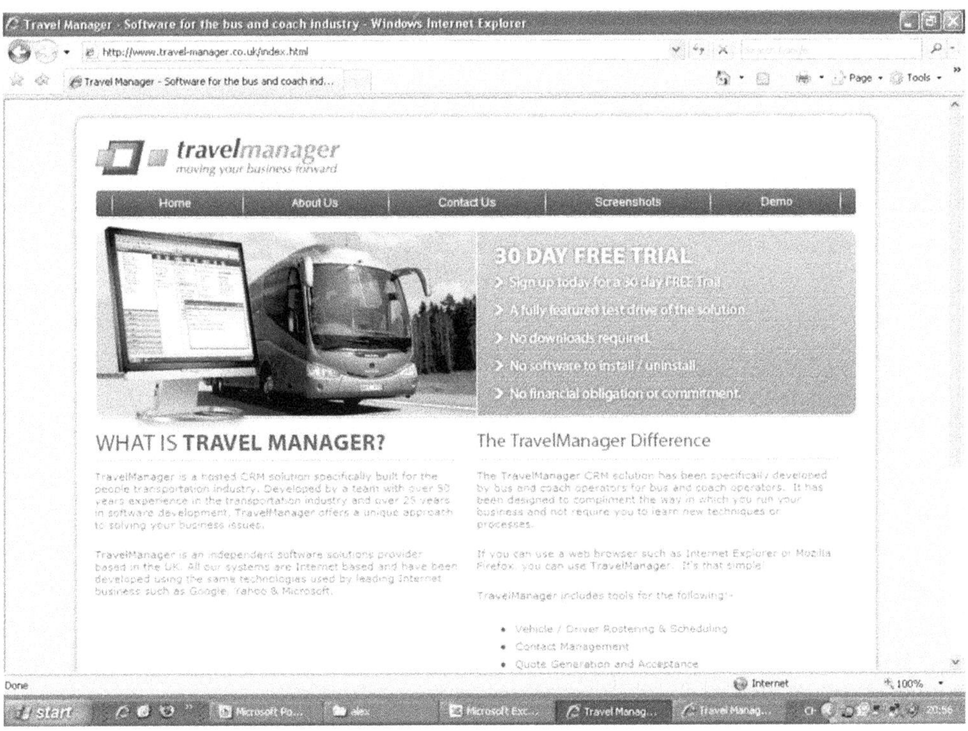

The new knowledge gained from the project significantly enhanced the company's ability to design, develop and market comprehensive web-based packages. The company thus "decided to close its traditional environmental polluting bus and coach business to become a competitive player in environmentally friendly IT centered e-business" (Technology Strategy Board, 2007, p.2). The project was the catalyst for a major change in company operations, and the coach and bus company was closed shortly after the end of the project. The son of the Beaumont Travel owner and managing director established Beaumont Business Software Ltd, based in Sheffield, to develop and sell business and domestic software. The company is still in operation today, but turnover is relatively small.

Analysis

The project at Beaumont Travel was given a B rating by the Technology Strategy Board, and the project scored highly on all the change factors (Figure 8). The transfer and onward development of new technology skills was central to the apparent success of the project, which allowed the transformation of this small company into a software house that enjoyed early success, and at one level, the project had few weaknesses. The Final Report (Technology Strategy Board, 2007) summarized the project thus:

We feel in every sense this partnership has been a huge success with an excellent outcome for all parties concerned. Contact between the partners was excellent. Without exception, both weekly supervision meetings and monthly project meetings were maintained regularly throughout the two years. The meetings proved to be very effective in maintaining focus and progressing ideas. Mr. Mike Sadler has directed the project internally, and chaired the local management committee. Mr. Dean Sadler (industrial supervisor) has provided close professional supervision and personal support for the Associate (p.2).

The project was led from the front by the company managing director; and his son - who developed his own technology skills in the course of the project - played a key role in the technology transfer process. The project was well aligned with - and the instigator of - the company's new business strategy. Requirements specification, software development tools selection and implementation execution were all well done, and project management capability and procedural discipline were of a high standard. However, in the longer term, the company did not develop as a major software supplier

Figure 8. Change factors in the Beaumont Travel software development project

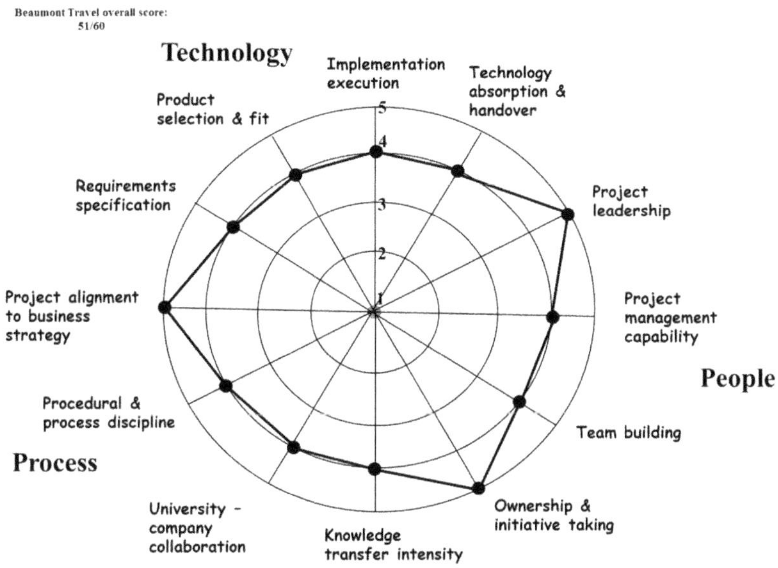

for the SME market as had been the initial aspiration; but within the context of the two year technology transfer project, the project can be viewed as an outstanding success.

NEW SERVICE MANAGEMENT SYSTEMS AND PROCESSES AT MUDDY BOOTS SOFTWARE, ROSS-ON-WYE

Company Profile

Muddy Boots Software Ltd (MBS) creates innovative software for the food supply chain, and in 2009/10 the company's business plan targeted a trebling of turnover from £1.6m to £6.0m in 2014/15. The company's business vision was to be a world recognized authority and trusted provider of traceability and quality assurance solutions for sustainable food and farming. The company's customers included Unilever, Marks and Spencer, Morrisons and Coles. UK operations were based at Ross-on-Wye in Herefordshire, but the company also had bases in Germany, Kenya, South Africa, Egypt, Australia and New Zealand. The company had three main software products that were the main drivers of revenue growth (Quickfire, Greenlight and CropWalker). A strategic component of the projected growth in revenues was to be the implementation of new customer centric systems and processes to drive and support new sales both in the UK and in overseas markets.

Project Objectives

When the project commenced in 2010, the company's software was deployed in over 40 countries and its user base spoke a range of different languages. The existing support systems and processes were inadequate to cope with this changing dynamic, and the overarching aim of the technology transfer project was to "develop new processes and methodologies to support the business growth going forward and enable it to provide the level of service required by its evolving customer base" (Technology Strategy Board, 2012, p.3). The company was transitioning from a mainly UK customer base to an international one, supported by additional offices abroad and support systems which needed to be accessible from multiple locations and different time zones.

Microsoft CRM was the key service management software in the company and was used alongside Microsoft Outlook with which it integrates. However, in addition, other departments in the company were using additional software packages for user support (e.g. Microsoft Team Foundation Server (TFS) for back office code management and a bespoke legacy system for timesheets), with obvious disadvantages – there was no common point for service management or one-business view for reporting. In a questionnaire survey of user perspectives conducted by the Associate, it was noted that "Microsoft CRM and Microsoft TFS have only recently been brought into use in this business and even though their capabilities are vast, they are not being fully utilized currently. Because service management is dispersed across several departments within the company, and several systems are currently in use, there are improvements to be made by integrating these systems more closely. Further work needs to be done to achieve this before the company grows further" (Tipton, 2011, p.1). More specifically, the project aimed to develop a technical framework that would enable improved software distribution and support for international systems operations, to specify, develop and implement new training systems to support sales operations, and to embed a new customer-centered support culture through intensive staff training and support.

Project Implementation

The project had five main stages (Table 3). Stage 1 involved the research of core business processes and development of current and future process maps (Figure 9) and an analysis of systems and technologies used at the time. From this analysis, stage 2 extended and improved the deployment of the Microsoft CRM system to allow the following functions:

- The system was used for all Muddy Boots products by international technical support teams and for software testing.
- Service Level Agreement (SLA) reports were held within the system and used with multiple customers.
- Bespoke reports were created for internal monitoring.
- Basics were in place for subsequent development of business intelligence reporting.
- The system could be accessed by Muddy Boots partner companies if required.

Table 3. Project stages and task list from Muddy Boots project proposal (Technology Strategy Board, 2010, pp.9-10)

Project Stage	Task List
1 – PROCESS MAPPING & SYSTEMS AND TECHNOLOGY REVIEW	1.1 Building company and industry knowledge
	1.2 –1.3 Process mapping: development and analysis
	1.4 Systems portfolio map
	1.5 Hardware review
2 – CRM EXTENSION AND ENHANCEMENT	2.1 Functional requirements specification
	2.2 – 2.3 Implementation
3 – NEW TRAINING AND E-LEARNING SYSTEMS	3.1 Systems Plan
	3.2 – 3.3 Review existing software packages
	3.4 Configure and install new software packages
4 – IMPLEMENTATION OF NEW TECHNICAL SUPPORT HUB	4.1 ITIL requirements, project brief development and submission to project board
	4.2 – 4.3 Implement new technical support unit
	4.4 – 4.5 Develop and review new technical support unit
	4.6 Develop extranet web portal
	4.7 Launch web based technical services function
5 – EMBEDDING OF NEW PROCEDURES & PROCESS CHANGE	5.1-5.3 User ownership, process embedding and systems handover

Figure 9. The new process map at Muddy Boots Software. The project spanned technical support, international operations, project management and sales

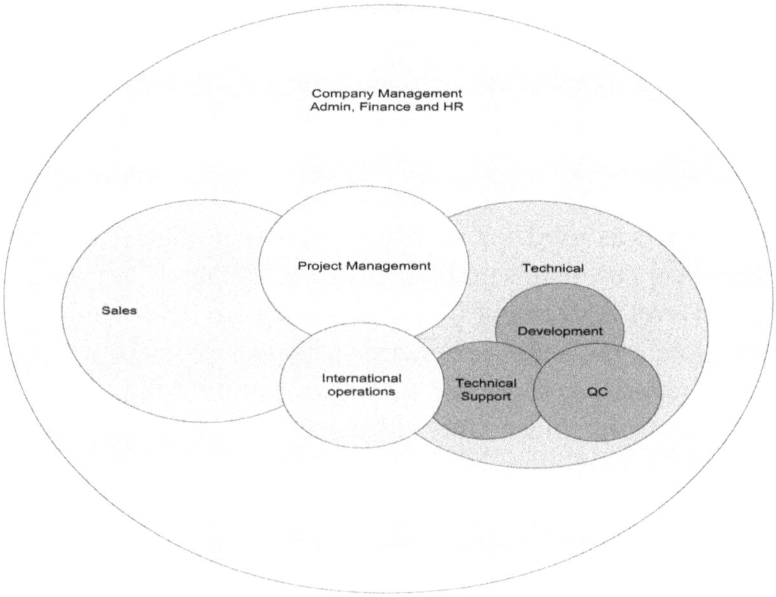

- The system could be used for user group review meetings with the company's technical services department.
- End-users were trained up to collect feedback on the system and contribute to subsequent improvements.

Stage 3 then focused on the specification, development and implementation of new training and e-learning systems. This involved the development of a range of video tutorials for the company's main software products to support both pre-sales and post-sales operations. In stages 4 and 5, the new technical support hub was set up for all international operations and efforts were made to embed a new customer centric support culture across Muddy Boots operations, based on the technical support future functions map developed in stage 1 (Figure 10). The Associate introduced the UK Government's ITIL methodology for service support (The Stationary Office, 2012), involving intensive staff training. Customer specific web portals were also established for on-going customer support. New training and support materials for the company's main products were developed and made available to both staff and customers.

Figure 10. The future functions map for the new technical support department

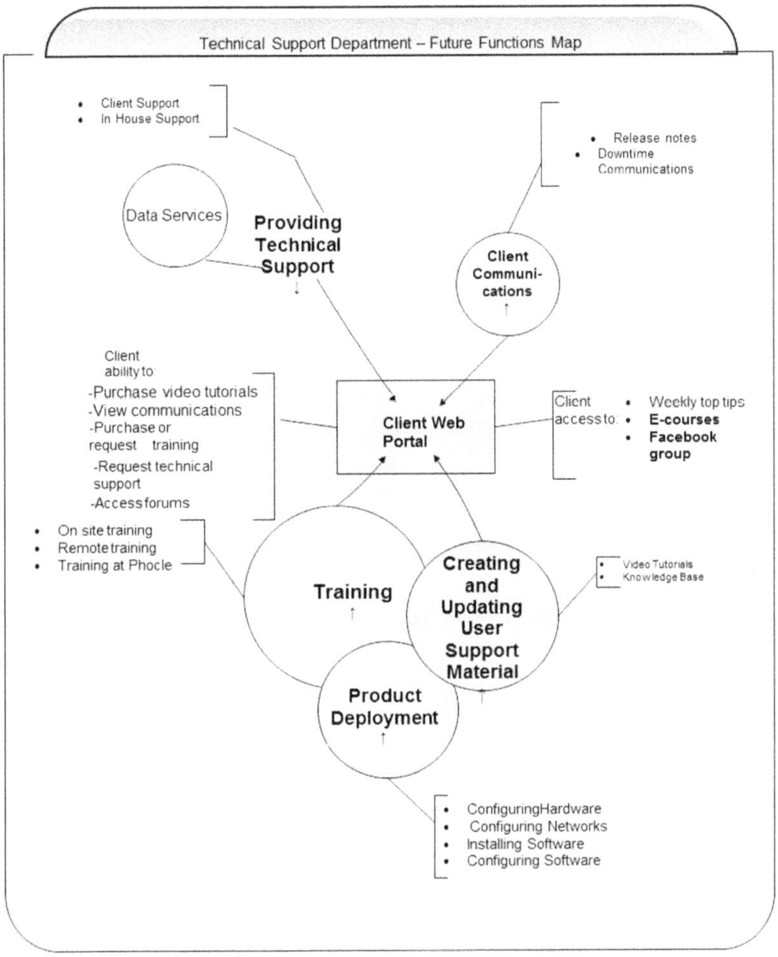

Project Outcomes

In the Final Report (Technology Strategy Board, 2012), the company reported that the project resulted in an "increase in capacity, function and knowledge of technical services function", and that it "upskilled the technical services division to provide a reliable, robust and effective function to the business" (p.4). Hitherto, a number of different software products had been used for service management activities which had been undertaken by different teams of staff, and the project succeeded in improving integration of both software products and associated processes. "The project has brought Technical Services into the business strategy, where previously it was seen as an IS function - it did not have specific attention from a strategic perspective" (p.4).

There had already been a lot of technically sound software in place, and the project provided additional analysis and a degree of re-implementation to optimize its deployment to support its expanding international operations. In addition, the Final Report (Technology Strategy Board, 2012) noted that "the company now has up to date user guides, tutorials, functional documentation for its products, and fast and effective means of capturing, storing and retrieving customer contact details and support cases. This has led to a more efficient technical support capability" (p.4). The project also enabled the company's management team to obtain a better overview of the technical services team's performance through advanced reporting from its systems and thus were able to identify pressure points and to gauge customer demand more accurately.

The company also reported that the new knowledge and capabilities acquired through the project had been successfully embedded into the business, for example, via the creation of process and workflow documentation, staff training, regular staff knowledge sharing workshops and management reports. The combined effects of the completed roll-out of the Microsoft CRM system, the development of a technical services portal, the upskilling of technical support staff, the introduction of documented process and procedures, video tutorials, the remote training capability, and the establishment of 24/7 support desk had all contributed to the securing of a number of high-profile accounts. This saw the company grow its turnover to £1.9m by the close of the project in 2011/12. The Associate, who had represented the company at an international conference and given a paper on service management (Wynn & Tipton, 2011), obtained a permanent position with the company as technical services support manager.

Analysis

The project at MBS generally met its objectives and was given a C grade by the Technology Strategy Board. The pre-existing range of systems were successfully rationalized, integrated and supported. New documentation and training materials were introduced. A new technical services support hub was set up and the Associate attained a new role on the company's payroll in the newly established department. The strengths of the project were that it was clearly aligned with the company's business strategy, and was led and supported by the managing director and other senior management.

Figure 11. Change factors in the Muddy Boots project

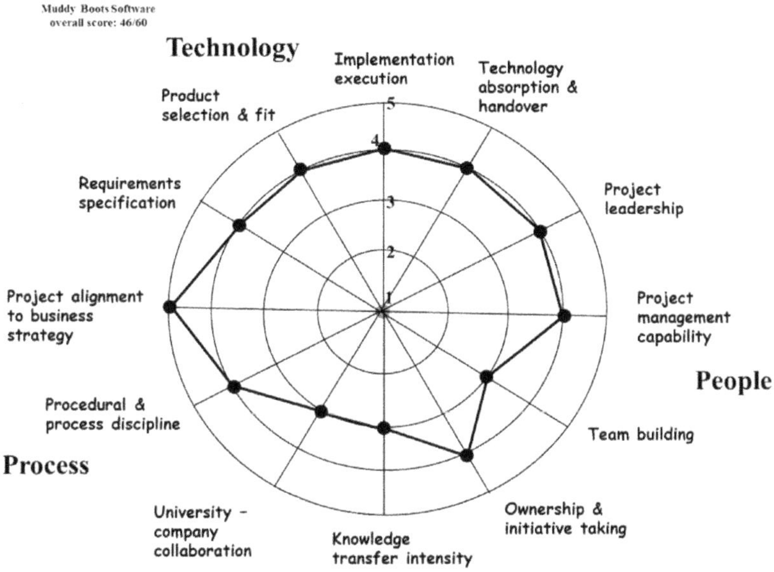

Project management and procedural discipline scored highly (Figure 11) underpinning an effective project implementation, but the project was only average on knowledge transfer intensity, team building and university-company collaboration. This was mainly because project direction and leadership came so strongly from within the company: the "pull" factors thus generally outweighed the "push" factors.

Nevertheless, in a wider context, the project made a significant contribution to the company's continued expansion and success. The managing director concluded that the "project has allowed us to support the delivery of new contracts secured, avoiding third party costs in servicing these requirements. This benefit will continue to grow as the business secures more high profile brands who require the level of service the company can now provide" (Technology Strategy Board, 2012, p.5).

CONCLUSION

Three of the four technology transfer projects studied in this chapter concerned new product development (Figure 12), reflecting the nature of the software industry. Tools are now available at no or low cost to develop

Figure 12. Software company projects: technology transfer focus and degree of innovation

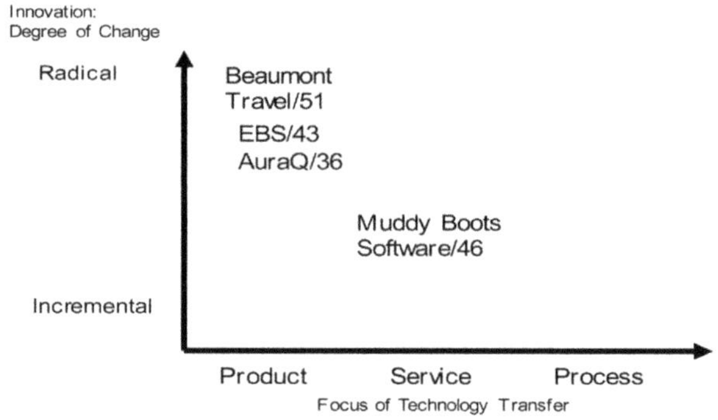

new products and thus radically change a company's products and profile. The challenge is to design products that offer something new and to attain both the technology and marketing skills to achieve success. For these three companies, the outcomes were mixed. At AuraQ, largely because of changes in the software marketplace, the new template products were never used, and at Beaumont Travel, after its transformation from transport to software house, the initial aspirations of becoming a significant player in the SME market never materialized. Only at EBS could the change of strategy and new product development be viewed a success, with major customers using their web-based products. All three projects were radical in what they attempted in terms of business transformation, but only at EBS did the project achieve its mid-term objectives.

The project at Muddy Boots was of a very different nature. It focused on service improvement for its growing international client base and was incremental in terms of the degree of change attempted. It was strongly aligned with its established business plan and firmly and effectively directed by its senior management, providing the infrastructure and process support for the company's continued growth.

REFERENCES

Infiniti Research Ltd. (2018). *5 Challenges That the Tech Industry Needs to Be Prepared For*. Retrieved September 14th, 2018 from https://www.businesswire.com/news/home/20180214005781/en/Major-Challenges-Faced-Tech-Industry-Infiniti-Research

Momenta. (2005). *Knowledge Transfer Partnership Project Proposal and Application Form Beaumont Travel*. TTI Database Record, KTP Programme Office. (Unpublished)

Technology Strategy Board. (2007). *Knowledge Transfer Partnership Final Report KTP 0878 (Beaumont Travel)*. Database Record. Technology Strategy Board. (Unpublished)

Technology Strategy Board. (2008a). *Knowledge Transfer Partnership Final Report KTP 1595*. Database Record. Technology Strategy Board. (Unpublished) Technology Strategy Board. (2010). Knowledge Transfer Partnership Project

Technology Strategy Board. (2008b). *Knowledge Transfer Partnership Associate Final Report KTP 1595*. Database Record. Technology Strategy Board. (Unpublished)

Technology Strategy Board. (2008c). *Knowledge Transfer Partnership Project Proposal and Application Form AuraQ*. Database Record. Technology Strategy Board. (Unpublished)

Technology Strategy Board. (2010). *Knowledge Transfer Partnership Project Proposal and Application Form Muddy Boots Software*. Database Record. Technology Strategy Board. (Unpublished)

Technology Strategy Board. (2011). *Knowledge Transfer Partnership Final Report KTP 7174 (AuraQ)*. Database Record. Technology Strategy Board. (Unpublished)

Technology Strategy Board. (2012). *Knowledge Transfer Partnership Final Report KTP 7223 (Muddy Boots Software)*. Database Record. Technology Strategy Board. (Unpublished)

The Stationary Office. (2012). *An Introductory Overview of ITIL® 2011*. Retrieved September 16th, 2018 from https://www.tsoshop.co.uk/gempdf/itSMF_An_Introductory_Overview_of_ITIL_V3.pdf (ITIL® is a registered trademark of AXELOS Limited, used under permission of AXELOS Limited).

Tipton, E. (2011). *Questionnaire feedback summary on the service management processes, related activities and systems support at Muddy Boots Software*. Muddy Boots Software Ltd. (Unpublished)

University of Gloucestershire (2007, July 3). *Presentation to Project Board/ Local Management Committee 5*. (Unpublished)

Wynn, M., & Tipton, E. (2011). The Deployment of Service Management Systems in SMEs – Three Case Studies. *Proceedings of Service Computation 2011 - The Third International Conferences on Advanced Service Computing*, 149-156. Retrieved September 16th, 2018 from http://eprints.glos.ac.uk/5452/

Wynn, M., & Zhang, S. (2008). *Innovation in the software industry: two case studies*. Paper presented at 8th International Business Research Conference, Dubai, UAE. Retrieved September 15th, 2018 from http://eprints.glos.ac.uk/5545/

KEY TERMS AND DEFINITIONS

.net (or .Net): A free open source product for use by software developers to build systems and applications for the PCs and servers, the web, mobile devices and games.

Business Process Management (BPM): An approach to defining and operating company business processes. Companies often use specialist software tools such as Metastorm to administer BPM, but in addition some Enterprise Resource Planning systems such as SAP have their own BPM tools for modelling, standardizing and automating processes.

Microsoft Team Foundation Server (TFS): A multi-faceted software product for the management of developer coding, encompassing the full product life cycle, from requirements management, product development, testing, and release.

SQL Server 2005: A relational database product released by Microsoft for storage and retrieval of data on a central server. There have been several subsequent releases of the product since 2005, most recently in 2017.

Chapter 4
Service Industry Companies

ABSTRACT

Ramaswamy (1996) suggested that companies that provide services normally conceive of a process as a sequence of activities needed to perform transactions that help to provide their services. To define processes is often problematic involving a range of complexities relating to customers, human behavior, and company structure. To organize the company around business processes, it is necessary to focus on external customers because business processes usually start and end with them. Processes have a line of activity which begins with an understanding and assessment of what the external customer wants and finishes with the external customer gaining what he or she needs and requests. The customer is always central within service organizations structured by process, when the key operational objective is to offer the customer more value in less time and with less cost. This chapter examines four small companies that attempted to do this through the introduction of new technologies.

BACKGROUND

The service industries often involve the provision of services to businesses as well as to final consumers. They are often termed the "tertiary sector of industry" in which no goods are produced. This is in contrast to "primary sector" industries such as mining and gas extraction, and "secondary sector" industries such as manufacturing and construction, where goods or materials of some sort are produced. This chapter examines the introduction of new technology into four service industry companies, one providing environmental consultancy services (Energist UK), another providing training services (C&G Services) and the third specializing in project management (Optimum Consultancy). Finally, the fourth project is at Allpay.net, a provider of financial services, particularly to the housing sector.

DOI: 10.4018/978-1-5225-7408-8.ch004

Service industries tend to be hit hard in times of recession, as their clients look for cost savings and review the use of third party services. The four companies suffered in the post-financial crash period of 2008 onwards, and this reduced their expected rate of growth following new technology implementation. Nevertheless, all have recovered well and continue to use the systems introduced in these projects.

NEW CRM SYSTEMS AND PROCESSES AT ENERGIST UK, CIRENCESTER

Company Background

Energist UK, based near Cirencester in Gloucestershire, is a small family business operating within the construction consultancy market. The company was founded in 2004 and had 40 employees in 2009, when the project started. The company provided assessment, certification, guidance, consultancy and training services to promote compliance with energy management legislation. A range of new regulations meant that the owners of buildings and factories had to adopt and deploy energy conserving materials, systems and services. These regulations included the 2006 Statutory Assessment Procedures, mandatory for all new build projects, and the Code for Sustainable Homes, which set out a series of increasingly stringent codes that were phased in during the ensuing years. The aim of the company was to become a leading player in a rapidly developing market for energy conservation services.

Project Rationale and Objectives

Energist UK had experienced rapid growth since it was founded, increasing turnover from just £156K in 2006/7 to £1.1m in 2009/10, but the company's systems infrastructure had not been developed or upgraded in an adequate manner to support this expansion. The company had a range of standalone systems that had been developed by end-users in spreadsheet and database software that were not integrated and contained different versions of key data items – particularly project and customer information. There was no clear technical strategy for systems development or technology procurement, and version control had also been lax, resulting in problems in software operation and maintenance. The lack of adequate systems integration and resultant data inconsistencies had affected management reporting across the company, especially at the customer interface. It had become difficult and time consuming to produce adequate reports on current markets and customers, and an implementation of a sound CRM system and associated procedures had become a top business priority.

With the introduction of new sustainability related legislation and regulations, the demand for Energist's products and services was likely to continue to grow, but would require a coherent IT strategy to guide the introduction of integrated, technically sound, scalable and well supported systems. The overarching project objective was to "implement new customer facing systems and processes to reduce costs, improve margins and drive new revenue streams to double turnover by 2012/13" (Technology Strategy Board, 2008a, p.1). The company wanted to make better use of their 8,000 customer profiles through the utilisation of a new CRM system. This aligned with their overall business strategy, which aimed to make their customers aware of all assessment needs regarding compliance with environmental regulations, and to support them in resolving these needs. Before the project got underway, the company had a fairly basic approach to customer relationship development, using Microsoft Outlook to store customer records within a range of shared folders. Reporting was limited in scope and data was often inconsistent, lost or duplicated, and the approach was not sustainable given the expected growth of the company.

Package Selection

When the project commenced in July 2009, and company management selected Microsoft CRM as the core software system around which to base their sales and marketing operations. They had previously trialled this product and as they already used Microsoft Office applications, an expansion of the software suite to include Microsoft's CRM software seemed logical. The initial software implementation was supported by a licensed distributor, chosen on the criteria of cost, capability and customer rapport. Specific modules were selected - they allowed a degree of configuration and yet, because of Microsoft's reputation and market position, were likely to remain technically robust and scalable. The modules selected were the sales module for data capture purposes (supporting customer management from sales lead to invoicing), and the marketing module, so that customers could be targeted and sold the correct services on the basis of increased customer knowledge.

Project Implementation

The project spanned a thirteen month period in 2009/2010 (Figure 1). The project was directed by the company IT manager, and managed on a day-to-day basis by the Associate. Third party consultants were also used during the implementation. Elements of the PRINCE2 project management methodology were used (Project Board, Project Initiation Document) combined with Microsoft's rapid application development (RAD) approach for software implementation. The Associate developed current (physical) and future (logical) data models (Figure 2) to guide the set up and configuration of the new system. The project progressed reasonably smoothly, although the requirement to parallel run new and old systems resulted in problems around data updates and staff resourcing. The relatively speedy implementation limited any overall negative impact. The implemented solution allowed integration with both Outlook and the company's Sage accounting package, although the Sage integration required the use of a middleware product (Sage Redware), which allows a 2-way data exchange between the two packages.

User groups were set up in the early stages of the project to establish new ways of working aligned with the new system and to discuss configuration options. The Associate ran regular workshops with these users to facilitate discussion and agree actions. The bedding in of new processes continued until the Associate left in late 2010 to take up a permanent software development position with another company. This was some months earlier than had originally been planned and impacted momentum in the final stages of the

Figure 1. The first two stages of the project plan at Energist UK 2009 (University of Gloucestershire, 2009, p.13)

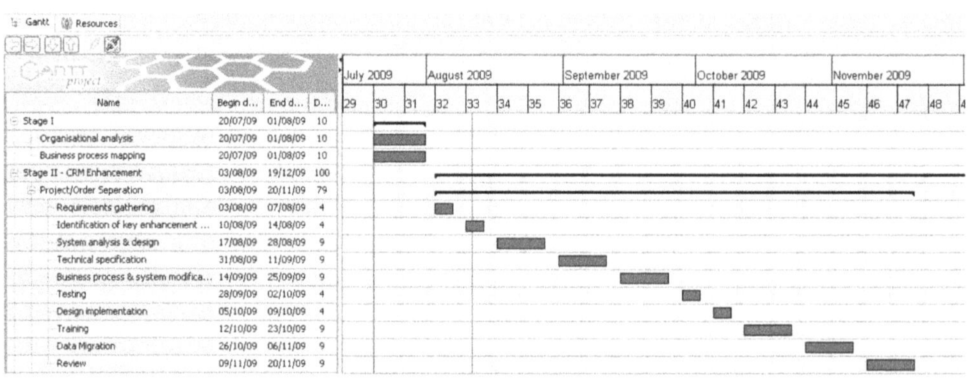

Figure 2. The logical data model for order capture in the CRM system at Energist UK (University of Gloucestershire, 2009, p.12)

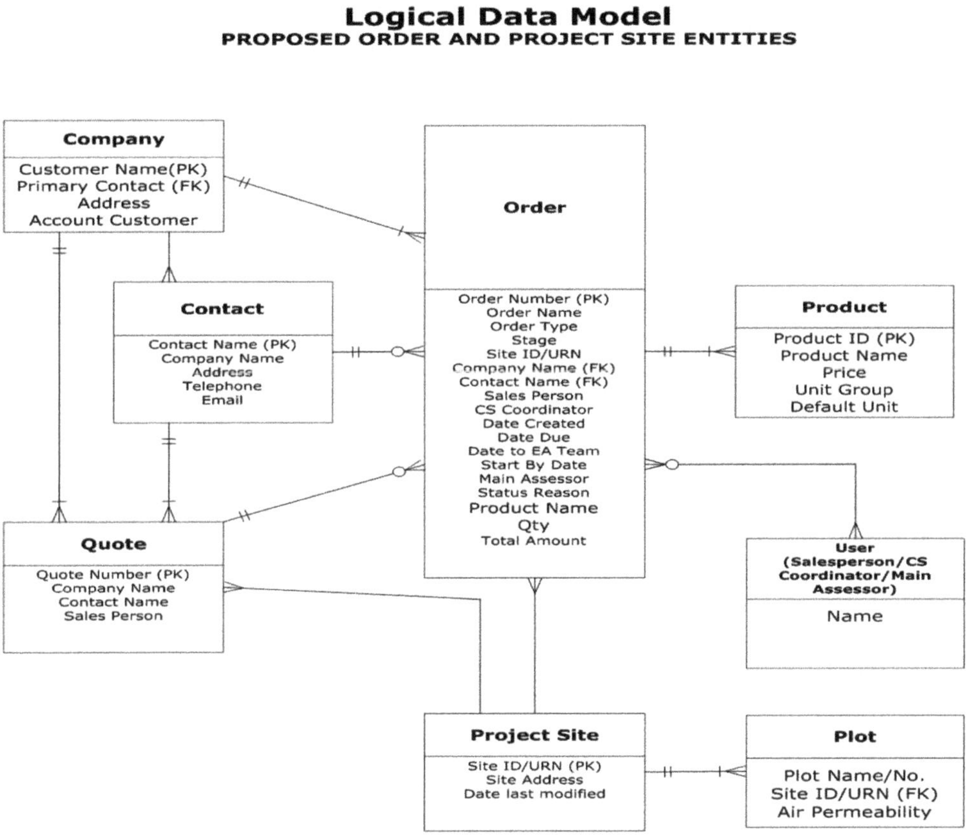

project. Nevertheless, the system was in place and had been configured to support new customer facing processes in the company, and the IT manager ensured continuity and on-going ownership.

Project Outcomes

The centralisation and consolidation of sales and marketing data, facilitating more effective marketing and customer profiling, was a major step forward for the company. The user-friendliness and configurability of the package allowed the system administrators and key users to enhance their core customer data with new details and concepts (data entities and attributes). Data entry screens and views of data were customised for individual users' requirements and job roles. Benefits included improved management reporting and better systems integration (in invoicing and servicing) as new versions of the package were released. The Final Report (Technology Strategy Board, 2010a) noted that the system allowed "much extended reporting, automated workflows and alerts" and "automated documentation from CRM – especially quote templates", and that the project had "established a test environment for staging future changes" (p.2).

The system is now well bedded in and upgrades to the new version of the package were undertaken in 2017. The project "helped to consolidate and manage information about our client base and the system now allows us to effectively market to, and communicate with, them" (Technology Strategy Board, 2010a, p.3). The system has supported the continued growth of the company with turnover increasing significantly since implementation in 2010/11 to reach £2.3m in 2012/13 and £3.0m by 2014/15.

Analysis

The three main objectives of the project were met: internal processes were changed to better support customer needs, notably in the sales and marketing functions; stronger customer relations followed the exploitation of the new system functions to respond more effectively to customer requirements; and increased company performance (turnover and profitability) was achieved. Nevertheless, the project was given a D grade by the Technology Strategy Board, and the change factors diagram (Figure 3) points out some relative strengths and weaknesses.

The project was clearly aligned with the company's business plan, the company IT manager provided strong project leadership, and overall project management capability was enhanced by the Associate's knowledge and experience, particularly in the initial stages of the project. However, the early departure of the Associate meant that the final stages of the project were undermined somewhat and overall the technology transfer was weakened. The company nevertheless had built up its own capacity for supporting the new system and for progressing process improvements, so the overall outcome for the company was positive. The company had chosen a product well suited to their requirements which was distributed and supported by a local third party. Despite the qualifications noted above - this was a well-planned, staged implementation of a mainstream software package, allied to measured improvement of their sales and marketing processes.

Figure 3. Change factors at Energist UK

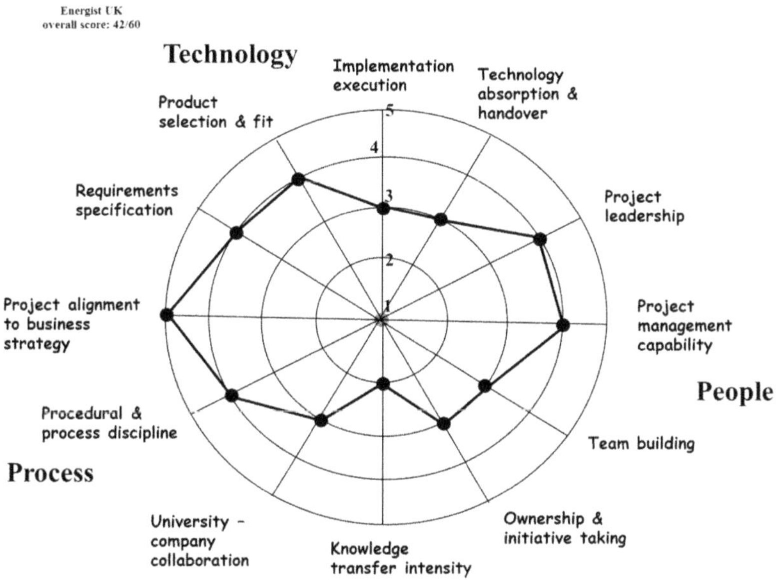

NEW TRAINING RECORDS SYSTEM AT C&G SERVICES, STONEHOUSE

Background

C&G Services (Europe) Ltd specialise in the provision of skills training to industries in the UK and the provision of health and safety consultancy services. The company generates revenues from a number of vertical markets, in particular utilities, plant manufacturers and machinery manufacturers. Prior to the start of the project, the company held several main contracts, including two with major utility companies - Severn Trent Water and Thames Water. The company had enjoyed steady growth for some years and was run and owned by husband and wife team Bob and Jane Oldmeadow, who were the main drivers behind the project that aimed to make business processes smarter and more efficient, allied to the introduction of new web-based software.

The company suffered from extremely detailed procedures that were put in place because of the need to maintain appropriate information for trainees using potentially dangerous equipment, so that the company were able to impart correct advice in any given situation. It was also believed that more modern systems would avoid the need for further administrative personnel. In 2003/4, the company had a turnover of £1.2m in a national market of many £billion turnover per annum.

Project Objectives

To support the continued growth of the company, business processes needed to become smarter and more efficient. The two-year project commenced in 2005 and offered the opportunity to implement new systems that would provide a state-of-the art service to major customers and engender more flexible and efficient ways of working. The overarching objective was "to improve existing business processes to create greater efficiencies and enhance company profitability through networked IT and web based systems focused on the customer" (Momenta, 2005a, p.1). This was to be achieved through the provision of readily available information and services over the web, such as access to delegate training histories, lists of available courses, the facility to complete booking forms online, and provision of feedback on the quality of training received. Without such an enhancement of customer service, the company was in danger of stagnating at the existent level of turnover and margin, without the step change in sales and profitability that the company management thought was achievable. The company was at risk of being "top heavy" in its ratio of administrators to trainers, and the technology transfer project aimed to address this in-house imbalance and at the same time improve service levels with major customers who were increasingly demanding new technology standards from their providers.

The old computer systems (Figure 4) had evolved over a period of years as the company expanded and took on new services and new staff. There were two legacy systems for course bookings, one developed in Lotus Approach and the other in Microsoft Access, and there was also a standalone Lotus Approach system used for purchasing. The Sage financials package was used for accounting and spreadsheets were used for analysis and reporting. Links between these systems were partly automated and partly manual. The level of IT expertise within the company was insufficient to develop systems that the largest clients were requiring. In addition, the company lacked any experience of business process remodelling and rationalisation that often precedes the development of new systems.

Figure 4. Pre-existing systems at C&G Services (Europe) 2005

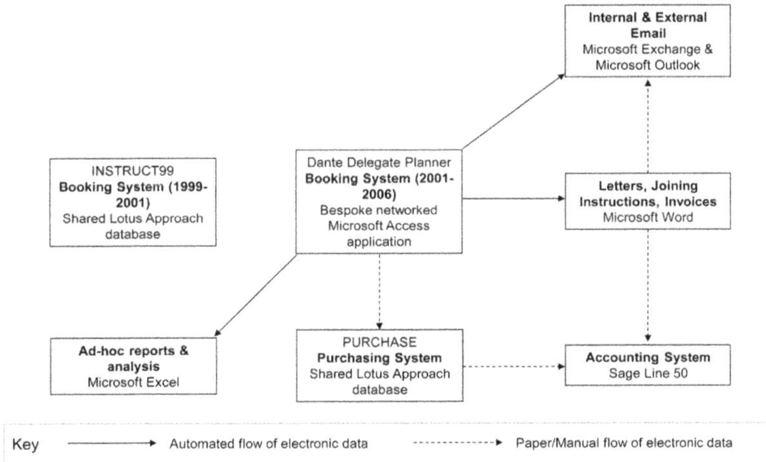

Project Implementation

In the initial stage of the project, the Associate researched and analysed the company's business processes and the systems and technologies used to support them. Then, together with the management of the company, a new administrative structure was formulated to accommodate significant process change to parallel the implementation of new systems. A full and detailed specification of the systems requirements for all main processes was carried out and the technical design was developed in liaison with the university supervisor, who had considerable experience in web portal development.

After a systematic review of available packaged software, the company elected to buy and implement the CS Course Booker product, a CRM and course booking system geared to the needs of the training industry. The system was set up and configured by the Associate in conjunction with the software vendor, and data was migrated from the old systems to maintain continuity with the new services being offered to existing customers. With the Course Booker software active for in-house staff, the web-based portal component of the system was configured and set up to allow access to information by staff working remotely and by major customers. The project was the catalyst for significant changes in working practices and attitudes to information and systems. Carefully selected staff members were trained as key users and systems administrators who took ownership of information provision both within the company and to external customers.

The new Course Booker system went live and was fully operational by late 2007, including the transfer of 5 years' worth of essential business data imported from legacy systems. The initial phase focused on the new course booking and administration functions, and the system was later made available to key customers. Expenditure on new systems and technologies totalled in excess of £50K. Although there was some in-house resistance to change on this scale, training the staff in the use of the new software provided greater awareness of what information was available and staff "are now enthusiastic to experiment with new and faster ways to achieve their tasks" (Technology Strategy Board, 2007, p.3). The Final Report also noted that the reporting facility in the Course Booker system "allows a greater opportunity to assess the performance of the company and individuals within it, and give detailed feedback for viability of individual services" (Technology Strategy Board, 2007, p.3).

Project Outcomes

The new systems portfolio (Figure 5) allowed the decommissioning of nearly all the old legacy systems. The new Course Booker software provided functionality to support sales, marketing, course booking and purchasing functions, and also had its own report generator. It also provided integration with the Sage financials package and Microsoft Office applications. Only one of the old legacy systems – a Lotus Approach database - was retained, and this was left as a standalone system to hold old data on course bookings in the event of customer enquiries.

Figure 5. The new systems portfolio implemented at C&G Services (Europe)

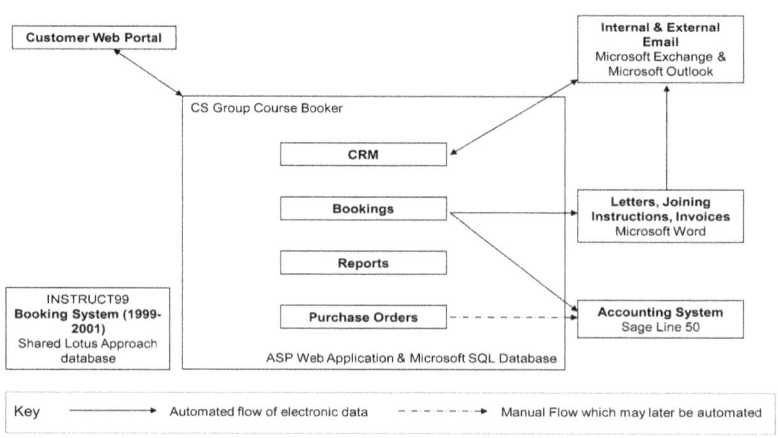

Faster information processing and better access to key data for clients and staff allowed more time for proactive work with existing and prospective clients. Greater information visibility gave the company improved knowledge regarding client refresher training schedules and associated requirements, allowing more efficient and effective management and allocation of company staff. The project was "the catalyst for significant changes in working practices and attitudes to information and systems" (Technology Strategy Board, 2007, p. 2). One example here was the scanning of course paperwork and its attachment to database records in the new system. The Final Report noted that "the company delivers in excess of 6000 candidate training courses per year and management of those records is vital for liability reasons. It commits to keeping these records for five years in hard copy. The scanning process eliminates the laborious task of locating hard copies when needed. In due course this process will be developed to fall in line with the legal requirements which will then allow the destruction of hard copies, allowing huge savings on storage and time and give direct access by the customer to this information" (Technology Strategy Board, 2007, p. 5).

The project was seen by clients in terms of improved and more proactive support facilities. The company was aware of the significant opportunities to win new business from the larger organisations it dealt with. The company targeted an increase its turnover of 20% per annum for the following two years, increasing that percentage thereafter as opportunities developed while at the same time holding back increases in administration costs. The expected net effect was that turnover would double from £1.4m in 2004/5 to £2.8m in 2009/10, and that profit margins would increase from 13% to circa 20% by 2009/10. Cost savings would come from a reduction in paper and storage of £15K per annum and headcount avoidance (as the company expanded) of four administrators across the company. As well as the major long-term benefits, there were a number of secondary benefits, for example improved IT support, upgraded network infrastructure, and enhanced IT knowledge. The company concluded that "the results of the project will put the company ahead of its competition and improve its profitability" (Technology Strategy Board, 2007, p.2).

Analysis

This project was assessed as a C grade by the Technology Strategy Board, and the change factors diagram (Figure 6) indicate some respective strengths and weaknesses. The project had many positives – it was closely aligned to

Figure 6. Change factors in the C&G Services (Europe) project

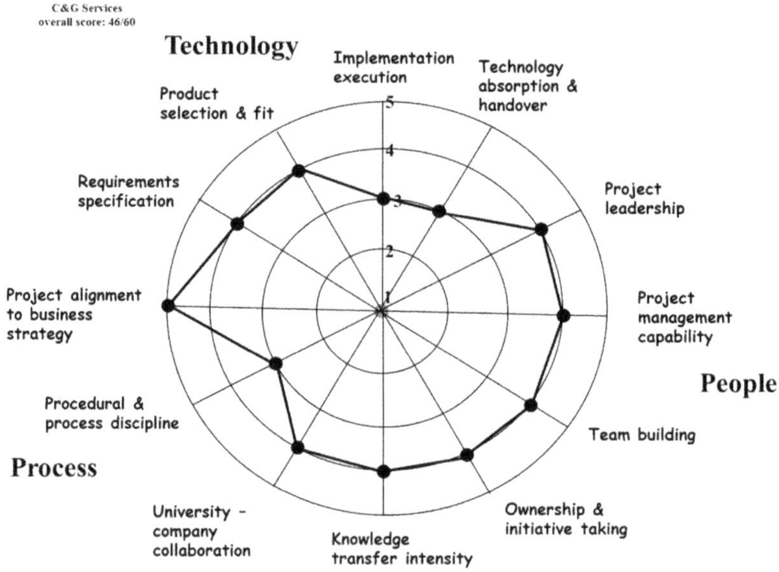

company strategy, it was well led and supported from within the company, team building and university-company collaboration were consistently sound and the project adhered to established project management principles and practice. The managing director chaired the project review meetings which helped ensure key issues were surfaced and managed, and the requirements specification and software package selection were carried out diligently and effectively. In this context, it is reassuring to know that the company is still using the Course Booker software in 2018.

However, implementation execution was not totally according to plan as delays were incurred, first in the software selection process, and later in data migration causing some months to be added to the original project plan. This put back the improvement in operations expected from the project, but once these issues were sorted out, company's administrative processes clearly benefitted. Course coordinators were able to carry through many more processes themselves, the new system providing them with a more comprehensive overview of the services offered to their clients.

This project highlighted the significance of information in the internal functioning of the company and at the customer interface. To adjust to this new way of working required in-depth training and new skillsets for company staff, producing a computer literate, customer-aware workforce. However, shortly after the end of the project, the financial crash of 2008 had a major impact on revenues, and the expected increases in turnover and profits did not materialise. Nevertheless, the company adjusted its strategy and kept trading, and in 2015/16 posted revenues of £1.6m.

NEW INFORMATION SYSTEMS AND PROCESS CHANGE AT OPTIMUM CONSULTANCY, CHELTENHAM

Company Profile

Optimum Consultancy, based in Cheltenham, Gloucestershire, was formed in 2008 through the merger of two companies - Hama Ltd, a project management services business, and J. Orchard Consulting Ltd, a surveying services business. In its first trading year (2008/9) the new company achieved a turnover of £2.4m and had 35 staff. Its core business was project and cost management in the property, engineering and construction fields, and its customer base at the time included major retailers, rail operators, major financial and banking corporations and sustainable developments. The technology transfer project focused on process innovation, achieved through the introduction of new information systems.

Project Objectives

Following the creation of the newly merged company in 2008, an internal review of systems and processes highlighted the challenge of combining and upgrading two different technology architectures and in particular the alignment and standardization of the sales processes across three offices in different locations (Cheltenham, London and Haywards Heath). The existing systems had been acquired and set up in an *ad hoc* manner whenever a need arose. Separate software and hardware systems had been purchased without detailed analysis of their longer term implications for IT strategy and process performance. Time lapses in distributing documentation constituted a significant risk to projects undertaken for customers, as there was no single source of project documentation which could be accessed by everyone in the company. The expanding network filing systems, delays in the sharing of project information, difficulties in finding the right electronic version of a document on the company networks, were all of major concern.

New integrated systems and customer facing processes were needed to provide infrastructure support for steady growth and improved margins, without the stop-start addition of administrative overheads. In addition, there was an urgent requirement to implement a refreshed and refocused business development strategy that crystallized the different roles of the company's three offices. The project was therefore focused on introducing new technology to effect process change. In terms of hard bottom-line benefits, the objectives were to reduce general administration time by 25% (mostly incurred in searching for documents and data re-keying), and thereby improve efficiencies in reporting, forecasting, monitoring and control. The main purpose of the project stated in the project proposal form was to "reengineer business processes and implement new collaboration and knowledge management software to drive a doubling of turnover and improve profit margins to 20% by 2011" (Technology Strategy Board, 2008b, p.1).

Project Implementation

The project was divided into distinct stages (Figure 7) and project managed using selected elements of PRINCE2, a mainstream project management methodology. This brought a disciplined approach to the acquisition and implementation of packaged business software. Initially, high-level business process mapping was carried out to develop a better understanding of

Figure 7. Project plan for the Optimum Consultancy systems project 2008-10

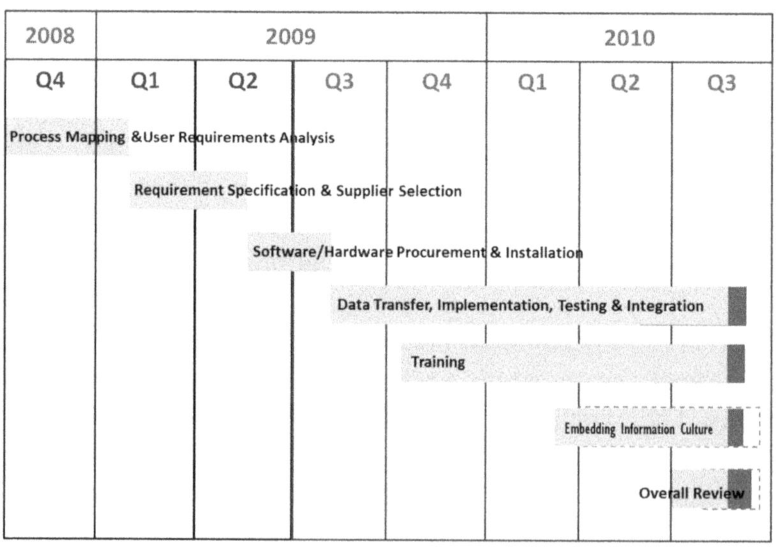

Optimum's core processes and to generate ideas for process improvement, highlighting information bottlenecks, data duplications and process delays or malfunctions. Analysis of the current systems used by the company was undertaken and overlain on a top level process map (Figure 8), which was used with key users from all three offices to establish current and future information needs.

This analysis revealed the disparate, non-integrated nature of existing systems (despite the fact that most processes were closely interrelated with each other). Existing workflows around the main systems still involved a significant amount of manual intervention. For example, the Sage software package was used by the finance department for sales order processing and invoicing, but it was not linked with systems used in the project management function.

After an evaluation of a number of packages available on the market, a web-based enterprise portal solution was recommended to the Project Board to provide an integrated collaborative environment for all business processes and to streamline the process flow across the three offices. The decision to proceed with off-the-shelf packaged software, implemented on on-premises servers, was taken and the suitability of collaboration software solutions

Figure 8. Process map and existing systems at Optimum Consultancy 2008

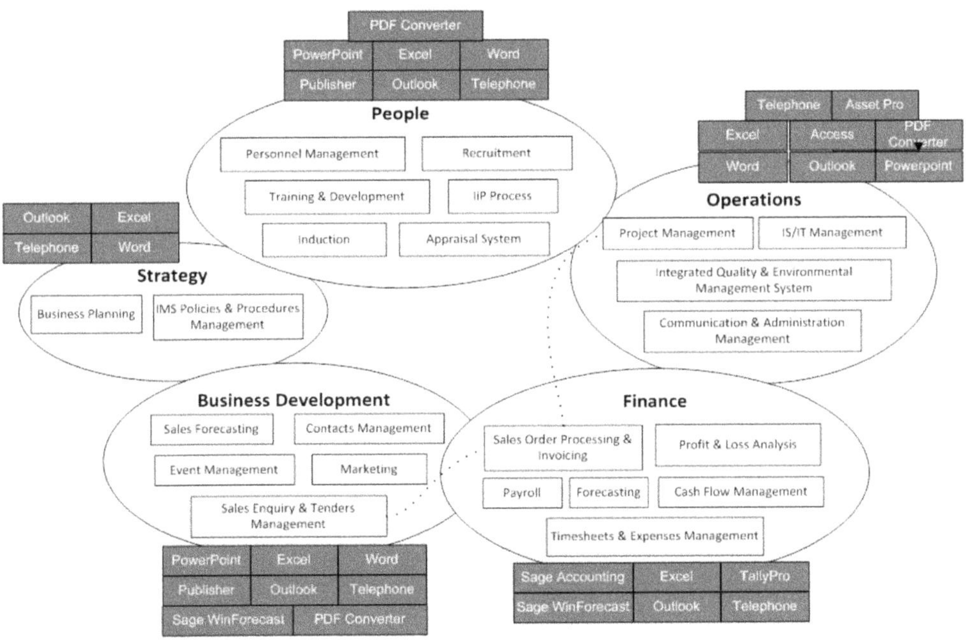

available in the market was assessed. The initial list of potential suppliers was compiled from current enterprise portal software available in the market, which encompassed the functions of document management, customer relationship management, and project and people management. The initial software supplier responses were evaluated using points scoring criteria and score card. From this evaluation, three suppliers with the highest scores were invited to a second round of evaluation, when a similar method was used to evaluate the suitability of software. At the end of this phase, the software package with the highest score was selected – Workspace from Union Square.

Implementation comprised two main phases. Phase 1 addressed the key business areas of document control, customer contacts and access to enquiry and project information; and phase 2 focused on integrating Optimum's finance operations into the Workspace software package. The company embarked on an internal communications exercise via company meetings and newsletters to ensure that everyone was aware of the project and its impacts and implications. Other activities during this period included unit testing of the main software modules, data migration of existing files and documents from the two old servers to the central server (Figure 9). Follow-up sessions with users were carried out periodically to reinforce training and ensure that the full benefits of the new software solution were delivered. The system was formally handed over to the company IT staff in June 2010.

Project Outcomes

The new systems allowed resource scheduling by field-based project managers, and the coordination of a range of activities carried out by architects, designers and surveyors. The project was a key enabler of continued growth and essential

Figure 9. Data migration process August - November 2009

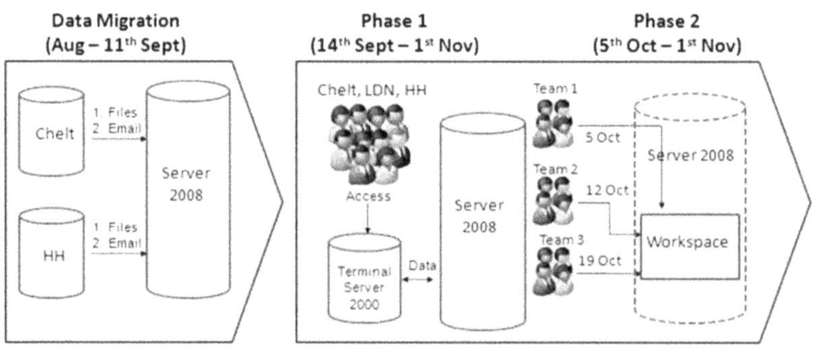

to the efficient operation of the company's core business. The lack of sound systems had meant administrative and management staff had been drafted in on an *ad hoc* basis to support the delivery of key projects. This produced fluctuations in turnover, profit and staffing levels.

The management of projects was made more efficient through the new technology. A bid having been won, all information was now ready to be automatically transferred to the project record, ensuring continuity and reducing errors; and the ability to find things more quickly proved increasingly useful as projects progressed. The new collaboration software integrated the management of time and resources and the recording of skills and training into the mainstream corporate database. Benefits included having instant access to forward schedules and availability without reliance on monthly paper/spreadsheet reports, which had proven very difficult and time-consuming to maintain in the past. The new system came to play a significant role in business development, keeping track of sales enquiries. Any work done on prospective jobs could be kept in the system. This allowed the senior management team to track and manage the new work pipeline more easily.

Quantifiable benefits came from a range of efficiency savings. "It is estimated that 5% of working time was previously wasted due to inefficient IT systems. Removing this waste contributes circa £60K per year savings. Avoidance of additional administrative headcount provides an additional saving of £35K per year per headcount saved" (Technology Strategy Board, 2010b, p.3). The ability to access information instantly about overall company performance, forward workload and future prospects as well as full details for every job, including who is looking after it, the client, fee type, value, allocated costs and the margin that is being achieved greatly improved the efficiency in all business activities. Based on the bottom-line benefits noted above (£95K p.a.), "the project has a payback period of just over one year, given the initial investment in software, hardware and staff of circa £110K in the two-year implementation period. This is in excess of original expectations" (Technology Strategy Board, 2010b, p.3). In addition, the project "provided the company partner with the experience and know how to effect cultural change in the organization. This is perhaps the most telling benefit long-term as it will help the company to continue to evolve and develop" (Technology Strategy Board, 2010b, p.3). Since the close of the project, the company has continued to be profitable, and in 2017, the Union Square software was still central to company operations.

Analysis

This project was given a B rating by the Technology Strategy Board, and it scored very well on nearly all change factors (Figure 10). The managing director provided strong project leadership from the outset, chairing the weekly project management meetings as well as the three monthly Project Board meetings with government sponsors. His direct involvement ensured the project was kept high on the business agenda and he provided critical inputs to decision making and problem resolution. This also provided an appropriate environment for initiative taking by project team members.

University-company collaboration was at the core of the project, and the university supervisor and company managing director worked on other projects, visiting company customer sites at Harrods and the Borough market in London to explore possible collaboration there. The intensity of technology transfer was also of significance as company staff had little experience of major software projects and the use of appropriate methodologies brought "a disciplined approach to the acquisition and implementation of packaged business software" (Technology Strategy Board, 2010b, p.2).

Some other key process factors stand out as having been of significance. Procedural and process discipline, allied to the Associate's knowledge of project management methodologies, was essential in keeping the project on

Figure 10. Change factors in the Optimum Consultancy project

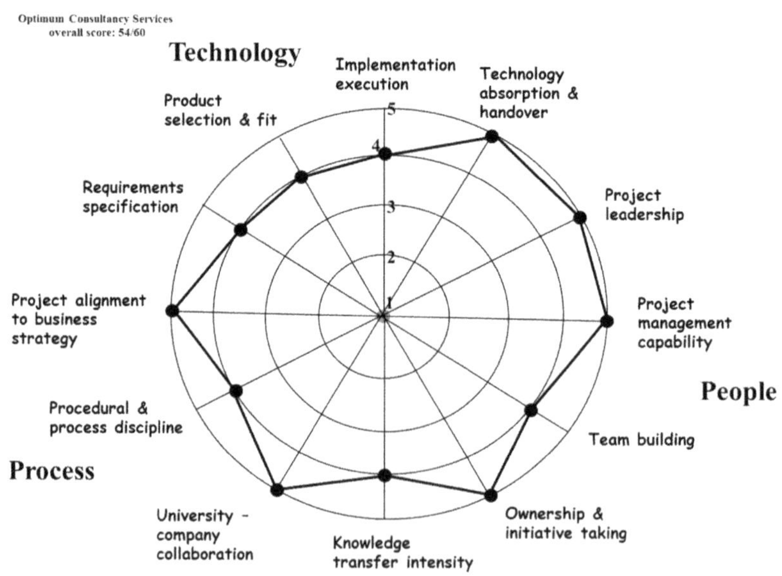

time and to budget. The alignment of the technology change with overall business strategy after the formation of the new company was also evident. The new systems were a key enabler of the successful merger of the two companies, enhancement of customer service and improvement in internal administration.

MIDDLEWARE DEVELOPMENT FOR SYSTEMS INTEGRATION AT ALLPAY.NET, WHITESTONE, HEREFORDSHIRE

Background

Allpay.net Ltd is a financial services SME, specializing in electronic bill payment solutions. Its main products are based on swipe card payments, internet payments and telephone payments. The company was founded in 1996 and experienced rapid growth in the following decade, building a reputable image within its marketplace and amongst its clients. In 2007, it was listed in the Sunday Times Tech Track 100. Turnover in the financial year to June 2008 was £21.8m and the company had 190 staff.

In 1996 Allpay.net started providing swipe card payment facilities for revenue collection for local government and housing associations. The founder and managing director had come up with the idea of electronic bill payments whilst collecting rents on behalf of the council. He found it frustrating working with paper based bill payment systems and concluded that technology could be used to speed up the process (Allpay.net, 2007). This resulted in greater choice for bill payers and a more efficient way of doing business for the clients. In 1996, the company (then named Financial Collection Services) had just two clients and four staff; by 2008, Allpay.net had over 650 public and private sector clients, and was one of the UK's fastest growing companies.

Project Rationale

The objectives of the original project as set out in 2005 were "to design, develop and market a suite of advanced software for the housing management sector encompassing central business management and payments collection modules" (Momenta, 2005b, p.2), but after a strategic review in 2006, the company changed its focus to try to improve internal systems integration. In

2007, when the project started, the company were using a number of bespoke systems that had both internal and external functions. These included a system for client account administration and reconciliation, a CRM system, an Intranet product (Nebula), and Purchase ledger, Banking and BACs transmission systems. In addition, the company had two web portals - termed "Webconnect" and "Allpayments" - which Allpay's clients used to access Allpay's services. Most systems were robust in functionality, but suffered from inadequate integration, with bespoke but limited interfaces, resulting from a lack of overall architecture planning and design. The maintenance burden for the IT team was disproportionately high, and, because of the need to meet tight development timescales, these systems were generally poorly documented, impeding their enhancement and scalability.

The project aimed to enhance the services that the company provided to its key customers, particularly the housing associations and local authorities. It focused on the provision of Payment Information Files (PIFs) to key customers, a process that had hitherto relied upon a plethora of bespoke files for specific customers. A thorough feasibility study confirmed that the in-house legacy systems that supported the PIFs could be replaced by a configurable, scalable middleware product. The legacy systems transferred and manipulated data in an unwieldy way that required much manual operation and intervention. Developing a new middleware application using up to date technology would reduce maintenance overheads and provide improved software architecture (Figure 11), which would help the company deliver a more robust customer service in the mid-term.

Project Implementation

The middleware development was an in-house software replacement project. The overarching objective was "developing a configurable middleware product to provide customer payment files to housing associations (and other) customers, and replace the old hard-coded legacy systems that had previously performed this function, which constituted significant maintenance and support overheads and provided poor customer service" (Technology Strategy Board, 2008c, p.2). The core project team consisted of three staff, with the Associate playing a hybrid role of project manager, business analyst and software developer. Given the nature of previous in-house software development, the team were determined to exert strong project management control to ensure product quality. It was decided to use selected elements of the PRINCE2 project management methodology. However, it was considered impractical and an administrative burden to use the whole set of PRINCE2 processes, components and techniques to manage the project. From the outset, the project team resolved to adapt the methodology to fit the project scope and scale. From the main PRINCE2 processes of "Starting Up a Project" (SU), "Initiating a Project" (IP) and "Planning" (PL), only a sub-set of the sub-processes were used.

In the very first project meeting, the Associate was appointed the Project Manager, and the Project Board (another PRINCE2 concept) was established, comprising the Project Manager, the IT Director (project executive and main user), the Systems Manager (main supplier) and University supervisor (quality assurance). Responsibilities were agreed and clearly and formally documented. A project feasibility study phase was added up front, in which it was reaffirmed that a valid Business Case existed for the project. A Project Brief was prepared, discussed and signed off, containing a formal definition of the project, highlighting its objectives and deliverables, scope, constraints and known risks, as well as formally identifying the project stakeholders. This was detailed enough to obviate the necessity of doing a Project Initiation Document (part of the PRINCE2 methodology) later on. Microsoft Project

Figure 11. Systems architecture for the new middleware product at Allpay.net

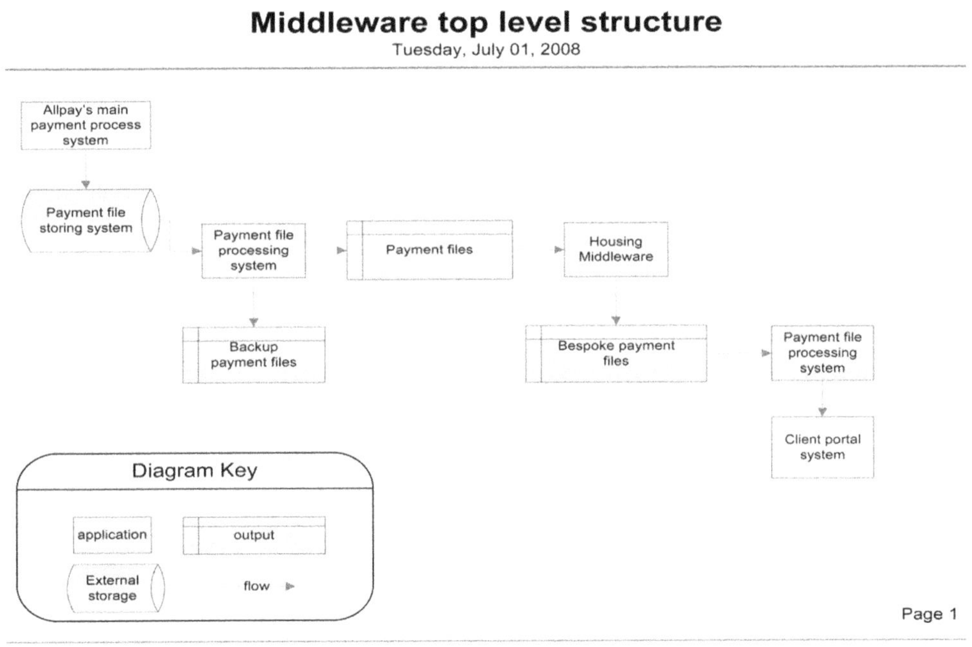

was chosen as the main tool to plan the project. The initial plan was drafted and the SU process completed.

Quality issues were discussed and documented in a published Quality Plan, which defined Quality Expectations, Acceptance Criteria, Quality Responsibilities, Standard, Control and Audit Processes and Tools. The project plan was revisited and published in more detail in Microsoft Project. As regards the Planning Process in PRINCE2, nearly all the elements were employed at some stage in the project-planning and re-planning was done throughout the project. The end product was defined and analyzed, and the team identified activities and dependencies, estimating and scheduling activities. The Business Case was refined and the Risk and Issues logs were set up and initial entries made. The plan itself was circulated and discussed at weekly project progress meetings.

Once underway, a major project in the PRINCE2 environment often goes through a number of stages which are managed via three main processes – Controlling a Stage (CS), Managing Stage Boundaries (SB) – i.e. at the end of one stage and start of another - and Managing Product Delivery (MP), which focuses on different products produced within each stage. However, in this project, once the feasibility study was signed off, there was only one main stage and thus the SB process became unnecessary; and the MP process was similarly not required as the main deliverable was the new piece of middleware - and resources, milestones and dependencies were planned and managed via the Microsoft Project plan.

The Project Board met weekly or fortnightly and undertook a range of management tasks contained in the Directing a Project (DP) process. Elements of the Closing a Project (CP) process were also used in July 2008 when the project was completed, which assisted the closure of the project in a controlled manner. Confirmation of project closure was discussed and accepted at the appropriate Project Board meeting. All deliverables were reported on, and the software application was released by the Quality Assurance team, who confirmed that the software met the specified Acceptance Criteria. The End Project Report was submitted to, and accepted by, the Project Board, and further developed by Project Board members in a Final Report submitted to the Technology Strategy Board. Additionally, meetings were held with the company IT Operations department and end users to discuss the system's deployment; and a further meeting was held with the in-house systems team to hand over documentation and agree support and maintenance arrangements.

Project Outcomes

The main project deliverables reflected the use of PRINCE2 and were stated in the Final Report (Technology Strategy Board, 2008c, p.3) as follows:

- Housing middleware project plan
- Project Brief
- Quality Plan
- Project risk and issues logs
- Housing middleware business requirements
- System requirements specification
- System design diagrams
- Middleware application systems development
- Acceptance testing/QA and user guide
- Technical documentation of the middleware product

The Final Report noted that "the project was key to implementing this new technical product and significantly enhancing customer service." (Technology Strategy Board, 2008c, p.2). The project illustrates how SMEs can use elements of a major project management methodology to effectively control and manage a project, which has been used as a point of reference by Bishop (2018) for developing a project management framework for SMEs. A summary check-list for the adaptation of the PRINCE2 methodology for IT projects in SMEs was presented at a conference in the course of the project (Wynn, Brandao, & Shen, 2008).

The Associate's Final Report (Technology Strategy Board, 2008d) noted that the project "illustrated how a tailored version of the methodology could provide a viable project management framework with a set of tools and processes that can be drawn upon as and when necessary" (p.3). In terms of hard benefits, the Final Report (Technology Strategy Board, 2008c) concluded that "Allpay.net now have a configurable middleware product to service payment file processing with leading clients. This will reduce overheads in the systems maintenance area by circa half a head a year (circa £20K p.a.), and lead to a significant improvement in customer service, particularly speed of implementation when new customers are engaged" (p.4). This meant that contract labour, that would otherwise have been required, was not necessary, saving the company an estimated £106K over the duration of the project.

Analysis

The Allpay.net project was well executed and delivered a complex middleware product that helped integrate the company's internal and customer facing systems. It was given a rating of B by the Technology Strategy Board. The Associate was a certified PRINCE2 project manager, and project management capability and procedural and process discipline were high (Figure 12). Although this was an internally focused IT project, it was nevertheless aligned to the company's business strategy. The Final Report noted that the project was "of relevance in the context of Allpay's business strategy, which is to significantly increase the size of the company over the next five years. The slick and efficient provision of payment information files to new and existing customers will be a key element of customer service" (Technology Strategy Board, 2008c, p.2). The project was given consistent support by the IT director of the day, but a high turnover of internal staff hampered team building somewhat over the course of the project. The company had three different IT directors during the project duration.

The Associate nevertheless exhibited good interpersonal skills, and university-company collaboration, ownership and initiative taking, and knowledge transfer intensity were all very positive. This was a 30 month

Figure 12. Change factors in the middleware project at Allpay.net

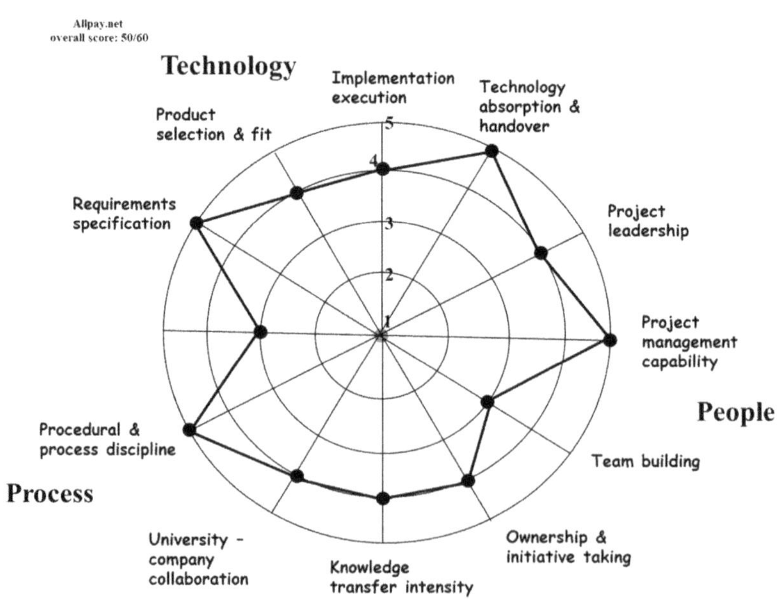

project and despite the high company staff turnover, the Associate developed good relationships with other IT staff that underpinned a successful systems handover. After the close of the project, she gained a post with the company as a business systems analyst.

CONCLUSION

The first three companies discussed in this chapter introduced new information systems and associated technologies in an attempt to improve their internal processes and thereby to improve customer service. The change introduced in these projects was generally incremental (Figure 13). At Energist, the project proposal noted that the "company urgently requires an infusion of knowledge to assess the best strategic forward direction for its IT/IS and the embedding of the technical skill set to implement the vision" (Technology Strategy Board, 2008a, p.3), and this was achieved through the implementation of additional Microsoft software in the sales and marketing functions, building upon the company's previous experience with other Microsoft products. At C&G Services, the "project was the catalyst for significant changes in working practices and attitudes to information and systems" (Technology Strategy Board, 2007, p.2), built around the new web-enabled course booking system. At Optimum, the change was arguably more radical as new systems and procedures were introduced to support the newly merged organization. The managing director noted that the project "was aimed at improving consistency of service and capacity rather than directly increasing sales. However, we have seen a 10% increase in turnover from £2.8m in 2008 to £3.1m in 2010 which can be directly attributed to the effect of the project" (Monk & Wynn, 2011, p.1).

Figure 13. Service industry companies: technology focus and degree of innovation

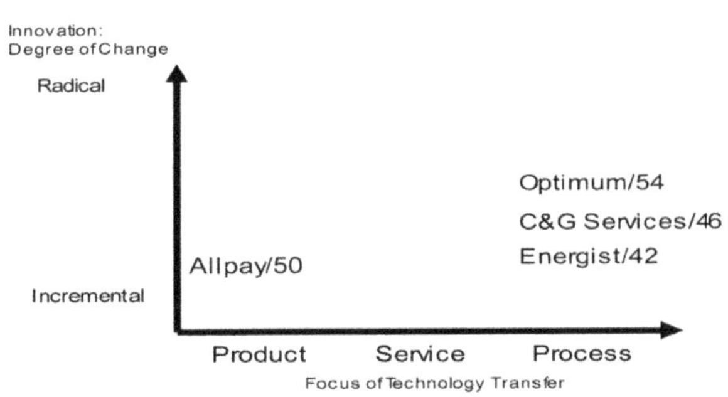

These three projects were similar in many regards. Optimum was arguably the most successful, and scored most highly on the change factors diagram. The role of the Associate in leading knowledge transfer and the embedding of new skills in the company was arguably the key differentiating factor. The fourth project at Allpay.net was different in that it used a mainstream project management methodology to develop a new product for internal systems integration, and in so doing helped improve customer service. The four projects also illustrate how an initial focus on process improvement or product development will, if the project is successful, also deliver improved customer service.

REFERENCES

Allpay.net. (2007). *The company*. Retrieved September 16th, 2018 from http://www.allpay.net

Bishop, W. (2018). A project management framework for small- and medium-sized entities: Accounting software implementation. *Journal of Economic and Financial Sciences, 11*(1). Retrieved September 16th, 2018 from https://jefjournal.org.za/index.php/jef/article/view/183/228

Momenta. (2005a). *Knowledge Transfer Partnership Project Proposal and Application Form C&G Services (Europe)*. TTI Database Record, KTP Programme Office. (Unpublished)

Momenta. (2005b). *Knowledge Transfer Partnership Project Proposal and Application Form Allpay.net*. TTI Database Record, KTP Programme Office. (Unpublished)

Monk, C., & Wynn, M. (2011). *Knowledge Transfer Partnership Case Study 5 - Optimum Consultancy*. Technology Strategy Board/University of Gloucestershire. Retrieved September 16th, 2018 from http://eprints.glos.ac.uk/

Ramaswamy, R. (1996). *Design and management of service processes*. Reading, MA: Addison Wesley.

Technology Strategy Board. (2007). *Knowledge Transfer Partnership Final Report (C&G Services)*. Database Record. Technology Strategy Board. (Unpublished)

Technology Strategy Board. (2008a). *Knowledge Transfer Partnership Project Proposal and Application Form Energist UK*. Database Record. Technology Strategy Board. (Unpublished)

Technology Strategy Board. (2008b). *Knowledge Transfer Partnership Project Proposal and Application Form Hama Ltd*. Database Record. Technology Strategy Board. (Unpublished)

Technology Strategy Board. (2008c). *Knowledge Transfer Partnership Final Report KTP 1156 (Allpay.net)*. Database Record. Technology Strategy Board. (Unpublished)

Technology Strategy Board. (2008d). *Knowledge Transfer Partnership Associate Final Report KTP 1156 (Allpay.net)*. Database Record. Technology Strategy Board. (Unpublished)

Technology Strategy Board. (2010a). *Knowledge Transfer Partnership Final Report KTP 7225 (Energist UK)*. Database Record. Technology Strategy Board. (Unpublished)

Technology Strategy Board. (2010b). *Knowledge Transfer Partnership Final Report KTP 6857 (Optimum)*. Database Record. Technology Strategy Board. (Unpublished)

University of Gloucestershire (2007). *Middleware Project Quality Plan (Allpay.net)*. (Unpublished)

University of Gloucestershire. (2009). *Project Initiation Document (Energist UK Microsoft CRM project)*. (Unpublished)

Wynn, M., Shen, R., & Brandao, R. (2008). *Managing IS projects in SMEs – Tailoring the PRINCE2 methodology*. Paper presented at Conference on Money, Economy and Management, World Business Institute, Imperial College London, London, UK. Retrieved September 16th, 2018 from http://eprints.glos.ac.uk/

KEY TERMS AND DEFINITIONS

Bankers Automated Clearing Services (BACS): The organization that manages and controls the clearing and settlement of UK automated payment methods, such Direct Debit and Bacs Direct Credit. It is now known as Bacs Payment Schemes Limited.

Middleware: Software that acts as a bridge to connect other information technology elements, such as applications, databases, web front-ends, normally across a network.

Rapid Application Development (RAD): A software development methodology that allows more rapid development than other methodologies. Sometimes called "Agile" development, it focuses on quickly developing the product and gaining user feedback, followed by iterative testing and redevelopment.

Chapter 5
Product Assembly, Sales, and Marketing

ABSTRACT

In developing and executing IS strategy, current business processes are analyzed, modified, or redefined in order to better support company strategic objectives and targets. Process analysis—an integral part of IS strategy development—can identify problems of information flow, data maintenance, systems integration, and process alignment with business strategy. Technology transfer projects in two product assembly companies are examined here. These projects developed and implemented IS strategies to align with contrasting business objectives: to provide stability and customer retention on the one hand and to provide key management information to support increased profitability on the other.

INTRODUCTION

In today's dynamic markets, organizations are using strategic information systems to help maximize their competitive advantage. Stadtler and Kilger (2008) pointed out that progress in information technology had opened up new perspectives for planning and controlling flows along a supply chain, and this is particularly relevant to those companies that assemble products from a range of suppliers and then sell them to customers further down the supply chain. The interrelationship between demand and supply chains was at the center of the initiatives in the 1990s to ensure better customer service through

DOI: 10.4018/978-1-5225-7408-8.ch005

more effective stock replenishment, based on new supply chain management concepts such as Efficient Consumer Response (ECR) and Vendor Managed Inventory (VMI). Lee and Ng (1998) discussed the two key approaches to improving the competitiveness of a supply chain - closer integration of the organizations involved and improved coordination of materials, information and financial flows. Van Goor (2001) has also concluded that "the integration of decisions within and between companies in a chain is the main mission. Information flows, financial flows and goods flows have to be integrated from a multi-company point of view" (p.3).

This chapter assesses three technology transfer projects in two product assembly companies. Two of these projects took place at TPG DisableAids, Hereford. The first project looked to develop, and then implement, a new IS strategy, and the second introduced new technologies into the customer facing processes to support the e-procurement requirements of some of the major public sector entities with which the company traded. The third project was at Beacons Business Interiors, near Brecon in Wales, where new bespoke systems were developed and implemented to interface existing software packages to provide new management information, notably regarding product profitability.

NEW IS STRATEGY AT TPG DISABLEAIDS, HEREFORD

Background

TPG DisableAids is a provider of equipment for the elderly and disabled. This is a "second-generation" family business, run by Alastair Gibbs (managing director) and his sister Amanda Harrold (finance director). The company is a value added distributor of a wide portfolio of products ranging from devices to allow arthritis suffers to safely utilise various household products and equipment, through to customised mobility equipment and patient lifting and hoisting equipment for healthcare professionals. TPG DisableAids also provides after sales support and maintenance for mobility, transport and lifting equipment to private and public sector customers. The company undertakes multi-site contracts to install, maintain and repair public sector equipment as well as providing disability equipment for entire healthcare facilities.

In 2005, when the project started, there were 1.4m people in the UK living in accommodation that required a stair lift, and there was clearly an opportunity for installing and maintaining these products more widely across the region.

In addition, the Disabilities Discrimination Act was forcing both public and private sector entities to install stair lifts and other equipment for the disabled. The company had less than 1% market share and yet had the logistical and operational capability to sell to a wider range of customers, including some outside their region. The company's business plan targeted a steady increase in turnover from £2.8m (in 2002/3) to £6.5m by 2009/10, mainly through organic growth in the region, notably with public sector entities. This required new business systems to integrate transaction processing, provide consistent management information and assist field operatives in their activities.

Project Rationale

The aim of the proposal was "to implement new computer systems to underpin the company's planned expansion over the next five years" (Momenta, 2004a, p. 2). Following the unsuccessful use of consultants in previous years, the company's senior management took the view that a new approach was required to reduce costs, increase productivity, ensure product traceability and provide enhanced service levels for an enlarged customer base. The company's information systems were centred on three main software packages – Sybiz Vision, Sybiz Vision Service Manager (VSM) and Sage Payroll. Sybiz was Australian owned and not supported in the UK very well, giving rise to upgrade, integration and performance problems. There were also multiple updates of key corporate data entities (notably customer and product data), which hampered the provision of key management information. The company had been trading for over twenty years and business processes had become well established, having been influenced by the capabilities and limitations of Sybiz Vision and VSM software. This resulted in silo-like processes that were self-contained and highly departmental and were tuned to the operation of the software rather than the most effective and efficient method for the company. In general the main financial, sales and order processing and inventory system (Vision) had sufficient management reporting facilities, but those of VSM were poor, being inaccurate or very basic, which were thus not useful indicators of performance either within their own right or in combined reports. The overarching objective of the project was to develop an IS strategy and then implement a set of new integrated systems.

Figure 1. Existing systems mapped to business processes at TPG DisableAids (Wynn et al., 2009)

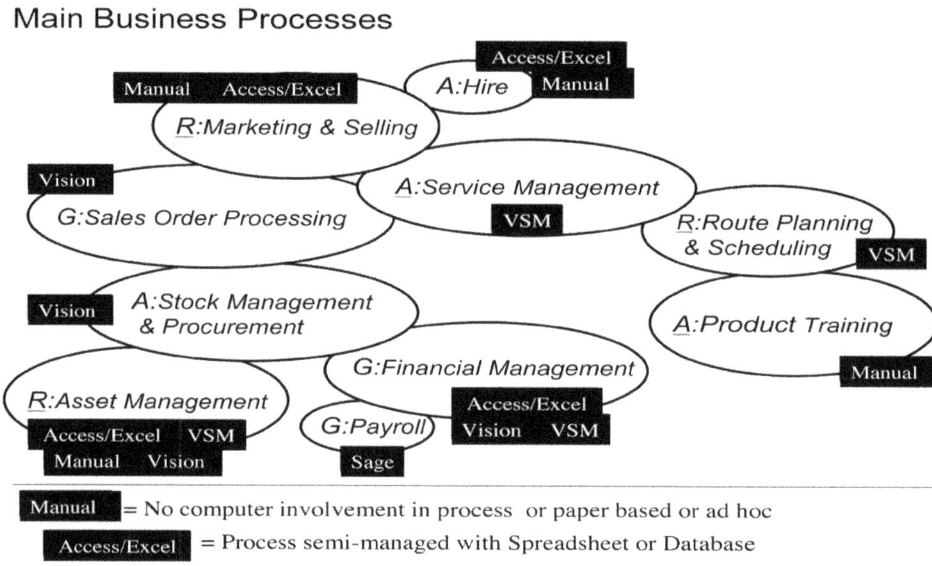

IS Strategy Development

This was a two year project comprising a number of distinct phases and stages that had been mapped out in the project proposal. Phase 1 focused on the development of a new IS strategy. An initial mapping of top-level business processes was followed by an assessment of existing systems and manual processes. Existing software systems were allocated to appropriate process areas and given a provisional assessment of red-amber-green (red being in definite need of replacement, green being acceptable in the mid-term and amber in need of further investigation - Figure 1). An analysis of the current needs of the business helped identify where new systems were needed and the Associate also investigated why management information was often contradictory, incomplete or impossible to generate (except manually). This analysis revealed that management information was deficient in most process areas of the company – only for sales order processing, financial management and payroll could it be considered adequate (Figure 2).

The Associate, working closely with the finance director, spent several weeks looking at options for software upgrades or total replacement. Implementing one integrated package - a full ERP solution - was seen as providing only marginal benefits given that 75% of the functionality of a

Figure 2. Missing systems functionality and management reporting shortfalls at TPG DisableAids (Wynn et al., 2009)

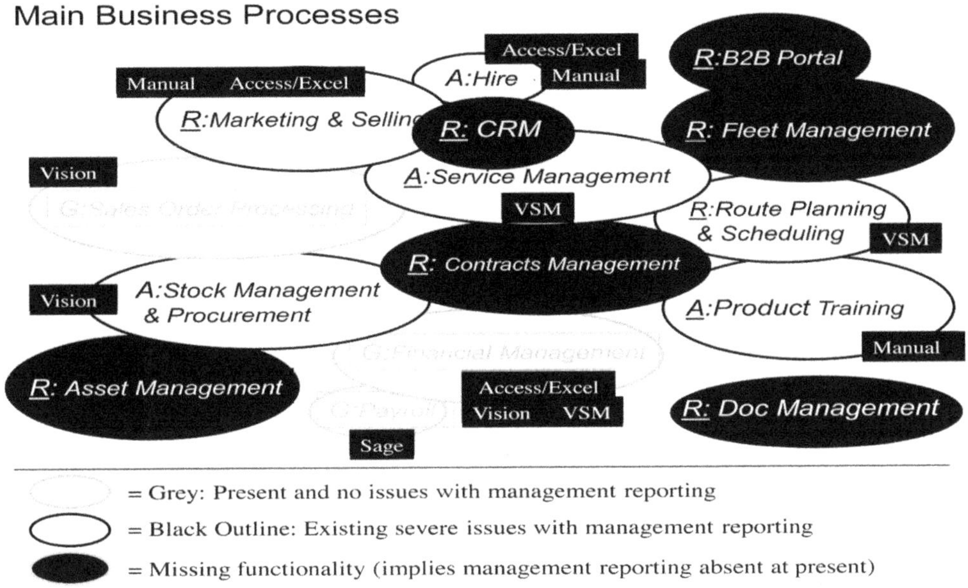

possible replacement system was currently provided in some fashion, and the likely resultant upheaval that an ERP implementation would cause. Consequently, point solutions and improved integration effort were seen as providing the best solution to meet pressing business requirements, whilst making good use of pre-existing software that was acceptable from a technical and strategic perspective.

In October 2006, three options were presented to the company Board. The first was to retain the current software, but upgrade these packages to the latest releases, cleanse data and improve report generation. This option was presented as being the cheapest short-term option but most costly in the long-term and also least likely to resolve change management issues. It was also doubtful if this would adequately support the company's plans for expansion. The second option was an extension of the first, with new additional systems for document management, customer relationship management, and fleet tracking, and associated process improvement in these areas of the business. This option was seen as more costly but of greater benefit in the short to mid-term; it would, however involve some change management issues that would have to be carefully managed through a phased implementation of new systems and processes.

The third option built on the second option but included a wholesale replacement of the VSM software package, which was used by the entire finance department for the monitoring of revenue and operating costs. Other financial and inventory systems would remain and be tightly integrated with the replacement product, which would require significant modification of existing business processes and affect the majority of the workforce in one way or another. This option would provide significant flexibility for future changes in business needs and would reduce operational costs, remove data entry, duplication and validation errors in multiple areas, and radically improve management. This option was the most expensive short term but provided the best options for future scalability and potential for incorporating and supporting business change. Although it presented significant change management issues, this option would provide a significant step-change in information reporting and management decision-making at all levels.

After considerable debate and a request for clarity on the respective business cases, the company Board elected to pursue the second option (Figure 3) on the basis of their intent to minimise investment at a time of uncertainty, and concern about the degree of process change required in staff managerial and computer skills. Plans were drawn up for investment and implementation of each additional package and the necessary time and tools required were estimated on an individual basis. This approach minimised risk of outcome, and retained firm control on investments as well as putting the onus on management and people rather than technological solutions to deliver a successful project outcome.

Figure 3. Selected software solutions at TPG DisableAids (Wynn et al., 2009)

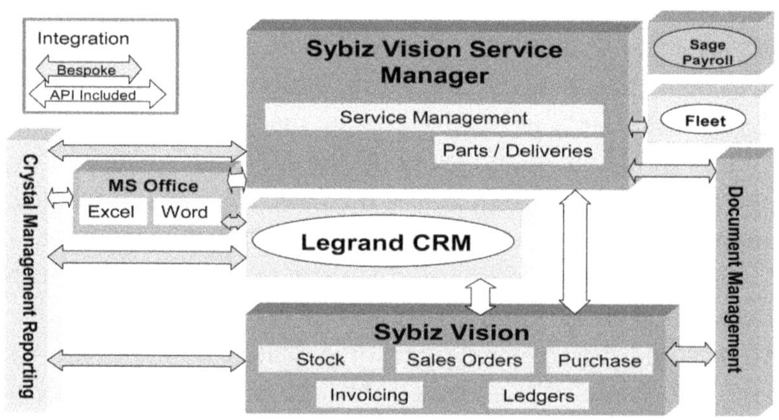

Strategy Implementation

The strategy implementation phase comprised a number of stages. Stage 1 established a stable technology platform and implemented some quick wins in the service department in the area of vehicle tracking and management of the fleet engineers. First, the Associate drew up a technology upgrade plan, gained budgetary approval and completed upgrading desktop hardware and software and a major network upgrade over a two month period. Desktop personal computers had been of variable standards, desktop software was at different version levels and there was no clear procurement policy for these products. Reliability of IT hardware and core and ancillary software was poor, and more than 66% of staff stated that their desktop computers hindered, rather than helped, them in their duties. Once these issues were resolved, in collaboration with the service manager and other senior management, vehicle management systems were investigated and evaluated, and a new system (Tracker) was piloted and then acquired, going live five months into the project. The engineers scheduling and tracking activities had relied on poorly informed, error prone and inefficient manual systems and mobile telephones that did not always work in remote areas. Implementing a new system to support these engineers was a significant achievement that gave the project momentum and a high profile within the company.

Stage 2 was the main systems implementation roll-out. The main elements, depicted in Figure 3, were:

- Upgrade of the Sybiz sales order processing, financials and service management software.
- Further configuration and training support for the new fleet management system (Tracker) with bespoke management reports.
- Introduction of a document management system providing basic scanning, document digitalisation and operational forms with search facilities, which allowed the re-deployment of office space previously used for document archiving.
- Pilot implementation of the Le Grand CRM system for order capture and integration with the Sybiz financials systems.
- Provision of new reporting tools across the systems portfolio for improved management information and to support process improvement and bespoke integration (using Crystal Reports and bespoke reporting over data cleansed using Artificial Intelligence techniques).

Stage 3 focused on the embedding of a new information culture and the management of a corporate training programme. The Associate organized and led a wide-ranging change programme that impacted in many areas across the business. In summary, these included:

- Further training and support of the main users of the Tracker vehicle management system, the upgraded Sybiz sales order processing software, and the 9 month pilot of the Le Grand CRM system.
- Introduction of personal and shared responsibility calendars and task list management for all personnel.
- Overhaul of the tendering process for large contracts, including new documentation and coordinated working practices across departments with new quality procedures.
- Day-to-day reinforcement of the value of IT to the business: its role in transforming error-prone and labour intensive paper-based processes by providing tools that were both supportive of individuals and the company's business objectives.

Project Outcomes

The project closed in 2007, having established a new IS strategy and implemented new systems in critical areas of the business to support its future growth aspirations. The Final Report (Technology Strategy Board, 2007) concluded that "at an operational level, the company now has the capabilities to use and onward develop a new range of applications that are supporting the company's sales processes and providing critical management information, notably in the cost estimating of projects. In addition, staff have been trained by the Associate to use the new systems and adopt new working methods" (p.4).

As a result of investment in technology infrastructure and new software, fleet engineers and sales teams were supported by the same number of administration staff that had not expanded in line with the growth in company turnover. The Final Report (Technology Strategy Board, 2007) stated that "a realistic estimate of saving is some £40K per year. Further benefits have come from the development of vehicle tracking reports that have highlighted critical issues in driver behavior and route planning" (p.4). The improved ability of the company to respond to customer requirements had numerous benefits in maintaining existing relationships and securing new business. This reduced the overhead on management in resolving data conflicts, particularly in the customer facing processes, and generally improved information quality and decision-making.

Across the duration of the project, the company continued to grow its turnover from £3.1m in 2004/5 to £3.9m in 2006/7. "This growth has been supported by improved information systems to allow better customer service and quicker response to customer enquiries. Moving forward, the company expects to gain further large contracts. This would not be possible without the supporting infrastructure of new information systems" (Technology Strategy Board, 2007, p.6).

Analysis

This project was given a B grade by the Technology Strategy Board and had many strengths as highlighted in the change factors diagram (Figure 4). The Associate was an experienced and knowledgeable IT professional who was able to bring a range of technical expertise into the company. Following the official end of the project, he was appointed the company's IT manager, ensuring the continued bedding in of new knowledge and skills. Ownership

Figure 4. Change factors in the TPG DisableAids information systems strategy project

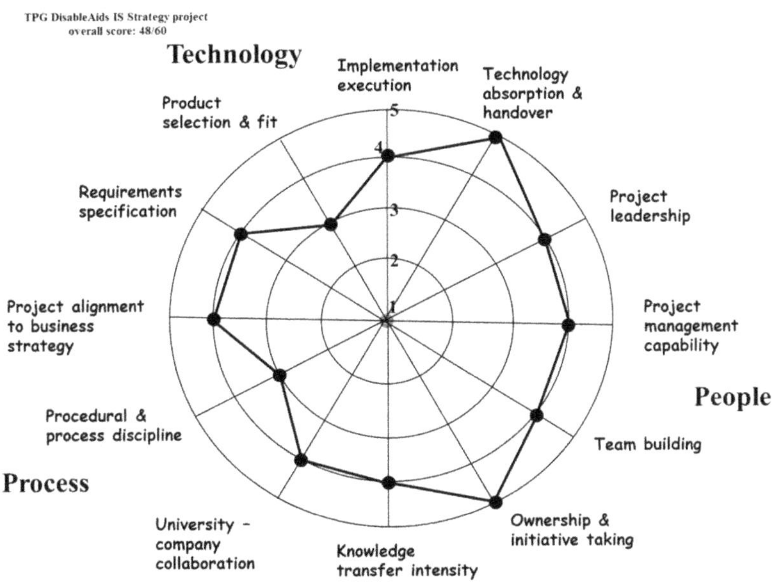

and initiative taking also stood out as a key strength and all round project management and procedural discipline were good.

IS strategy development was conducted in a logical and structured manner, and the decision of the company Board to pursue a conservative, incremental change strategy worked out well as the Sybiz Vision and VSM software packages are still in operation and supporting financial and service management operations in 2018. It is possible to argue that the software package selection was sub-optimal, but with the financial crash and ensuing recession closely following project closure, turnover targets were scaled back, and in 2014/15, the company had grown steadily but modestly to annual revenues of £4.5m. The election to pursue a low risk, moderate investment strategy was actually well suited to the broader business environment of the past decade. Nevertheless, under the guidance of the IT manager, IS strategy is regularly re-assessed, and the replacement of the legacy systems remains under review.

BESPOKE WORKFLOW DEVELOPMENTS AT BEACONS BUSINESS INTERIORS, BRECON, WALES

Background

Beacons Business Interiors (BBI) was established in 1989 as an interiors construction company providing its clients with facilities support services. When the company embarked on this project in 2004, it had 45 employees, a turnover of £7m and operated from bases in Brecon, Leeds and Cheltenham. It had steadily increased its turnover by focusing on "blue chip" clients, incorporating office furniture installation and design, mechanical and electrical systems design and related construction services. The facilities support market can be divided into different segments - blue chip, the SME market and the servicing market. In total, the market was valued then at £1billion, and BBI was in a dominant regional position within Wales, with only a small number of competitors. These were similar sized organizations, which offered comparable services, but without the same level of internal resources. Nationally, the competition was more comprehensive in a highly fragmented market place.

Before the start of the project, BBI had implemented some new core information systems (the Sage financials and the Goldmine CRM packages), but the expected bottom-line benefits of this investment had not been fully realized. BBI needed to rationalize its business processes and improve information availability and management control to drive company growth. Its business plan targeted £15m turnover by 2008/9.

Project Objectives

The project proposal stated that the aim of the project was "to improve communications and information availability across the company through exploitation of 'state of the art' workflow, communications and data warehouse technologies" and also noted that "to support and drive through planned company growth, new systems are urgently required to speed communication and improve information provision" (Momenta, 2004b, p.2). Senior management realized that BBI needed to rationalize its business processes to take advantage of existing information systems, and then to leverage additional benefits through the implementation of new communications and software integration technologies. The project involved the development and implementation of new systems and interfaces to improve management information and support efficiency gains, and the remodeling of corporate business processes. The project also established key performance indicators (KPIs) to assess the success of the change programme, and carried out cross company training in support of the new systems and associated information culture.

Project Implementation

The project proposal set out a two year outline plan (Figure 5), and following an initial assessment of existing business systems by the Associate, a two-stage approach to project implementation was adopted. The Associate acted as the company IT project manager and worked in conjunction with the company's senior management to plan and implement the project. Stage 1 provided some "quick win" improvements in software integration and provision of new management information, particularly in the areas of profit analysis and sales reporting. New process maps were produced encompassing improvements to certain business processes, and KPIs for success of the change programme were established. The main impact was in the areas of procurement, invoicing and project costing, where processes were simplified and supported by new management reports that avoided data duplication and manual calculations.

Figure 5. The outline plan for the 24 month BBI project (Momenta, 2004b, p.40)

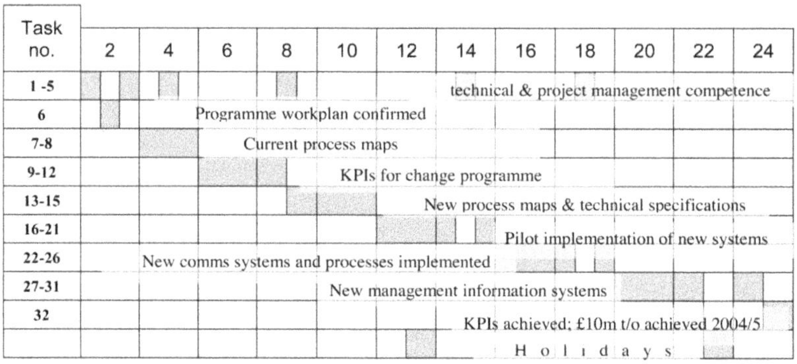

Stage 2 took a longer-term view of the systems that were required for the company to support its projected growth over the next five to ten years. The "quick wins" generated in stage 1 in themselves proved effective enough to support a significant growth in sales turnover, culminating in winning a £1.75m contract with the Office for National Statistics in November 2004. The main improvements in management information introduced during the project are listed in Table 1. Some of the key benefits came from the

Table 1. Key efficiency gains from new information systems at BBI (Wynn & Jones, 2017, p.7)

	Before the Project	After the Project
1	The company received about 30 invoices a day. To register these invoices and channel them to the right person was taking about 2 hours a day.	Using the new Project Identifier Program (PIP) this process took only ½ hour a day, a major reduction.
2	The project-costing process took 1 hour every day to process all the invoices and one full day every month to make the necessary calculations and produce the project-costing sheet. These sheets were neither accurate nor on time, as the recording and calculation were done manually.	A new Project-Costing Sheet (PCS) meant that project-costing could be done automatically. Manual calculations were no longer necessary, allowing timely profit analysis by customer and project.
3	For managing deliveries and suppliers, the project support team had to use one database to find out the purchase order number and then use that code to obtain the purchase order details from a second database. This was time consuming and error prone.	Using new programs written in Visual Basic and Excel - Purchase Order Processing (POP) and Sales Order Processing (SOP) - purchase orders, the supplier details and delivery status were organized on one sheet per project. As this was an Excel sheet, it also gave the project support team the facility to search quickly for any item of information that they might need, which was related to their project.
4	It took the accounting department about a day every month to sort out the site staff wages.	Using the new Time Sheet Database (TSD), this was reduced by 75% and could be done in two hours.
5	Site staff performance management was not effective as there was no electronic database to make the necessary analysis.	The TSD was equipped with the necessary analysis, making the management of site staff performance more effective.

development of an integrated project-costing sheet, which eliminated the re-keying of data and provided an on-line snapshot of project profitability. This gave the estimation department a quick effective feedback tool to guide their pricing and eliminated painstaking manual work in working out contract detail. A further enhancement was the introduction of automated programs (POP and SOP) to produce on-line reports for all purchase orders and sales orders related to a specific project.

This was of particular value to a company that adds value through design, acquisition, assembly and sales, and is something that neither the Sage Accounting nor Goldmine customer database systems could easily provide. BBI's procurement department would record purchase orders and sales orders for a specific project in the customer database and then track details from Sage, as a separate operation. The automated program, written and implemented by the Associate, linked the two systems and provided one automated project report. Another program provided detailed sales analysis by project and customer, again pooling data from non-integrated systems to provide improved management information and control.

This policy of delivering cost effective links and management reports (using Visual Basic Applications) continued right throughout the project, producing a wide range of "mini-systems" and reports in many parts of the business, building on the mainstream software packages implemented at BBI (Goldmine CRM and Sage Accounting). Stage 2 also produced a set of options for a longer-term implementation of a full ERP system (Table 2) that the company would keep under review as it continued to expand. The project

Table 2. Longer-term IS options at BBI (University of Gloucestershire, 2005, p.5)

	ERP Systems (or point solutions)	Capital Cost (£K)	Annual Support (£K)	Users (no. of licences)	Modules Included
1	Access Accounts & Goldmine	37	3.0	Main: 6 L. @ £500 CRM: GM Existing Licences.	Accounting, Finance, Project Management, Project Costing, CRM
2	SAP Business One	45	4.3	Main: 6 L. @ £1020 CRM: 40 L. @ £306	Accounting, Finance, Project Management, Project Costing
3	Sage Line 200 & Sage CRM	79	7.8	Main: 6 @ £1710 CRM: 40 L. @ £550	Accounting, Finance, Project Management, Project Costing.
4	Microsoft Navision	105	13.9	Main: 6 L. @ £1750 CRM: 40 L. @ £704	Accounting, Finance, Project Management, Project Costing, Gantt Chart Facility.
5	Microsoft Great Plains	109	14.4	Main: 6 L. @ £1150 CRM: 40 L. @ £797	Accounting, Finance, Project Management, Project Costing, Gantt Chart Facility.

concluded with a number of initiatives to embed the new information culture in the company. These included training sessions run by the Associate in a number of areas, including:

- Training to consolidate the changes made in internal processes, and to ensure that best use was made of the new information reporting systems.
- Training to allow on-going support and maintenance of new systems in Visual Basic and Excel. This training was for end-users who had become adept in using the new systems but also for a new information support analyst who had joined the company.

Project Outcomes

The ability of BBI to respond to customers' information and collaboration requirements had numerous benefits in maintaining existing relationships and securing new business. The introduction of comprehensive, integrated systems solutions, which shared a common database, helped to minimize data duplication and data errors. This reduced the overheads on management to resolve data conflicts and improved information quality and engendered effective decision-making. Another benefit to BBI was to allow flexible working from home. Many of the company's staff were away from the office for much of the working day. Hitherto, effective access to key information from remote PCs and laptops had been impractical because of lack of systems integration. This had frustrated attempts to encourage remote working by sales staff and engineers. The new systems platform made this possible. The Final Report (Momenta, 2006) noted that "the company now has the capabilities to use, support and onward develop a wide range of new applications that are providing critical management information, notably in the costing of projects and profit analysis." (p.4). In addition, Simon Oldroyd, the company Commercial Director, commented that "already we are seeing significant improvements in information availability that supports our drive to increase turnover and bottom-line profitability", and that the project was "central to the delivery of our five year business plan" (Monk & Wynn, 2007, p. 2).

Analysis

The project at BBI provided the company with better integrated systems, significantly improved management information, and some new skillsets and awareness of how to build links between disparate systems. The project received a B grade from the UK Government's funding agency, and scored moderately or well on all the change factors (Figure 6). It was well led and supported by the company's senior management and adequate in terms of project management discipline. The Associate took the initiative in finding interim solutions to link systems together and provide improved management information, but these solutions were so effective that they became more than short-term fixes. This led to some confusion as regards the overall direction and strategy, but issues were discussed and ironed out effectively in the project management meetings. Team building was only moderate and knowledge transfer intensity was limited, despite the training sessions organized by the Associate. The longer-term options for a wholesale replacement of IS which had been evaluated by the Associate were not pursued; but nevertheless, the transfer of new technologies into the company supported its planned expansion reasonably well, protected its existing investment in software packages and required relatively little new capital investment.

Figure 6. Change factors in the BBI project

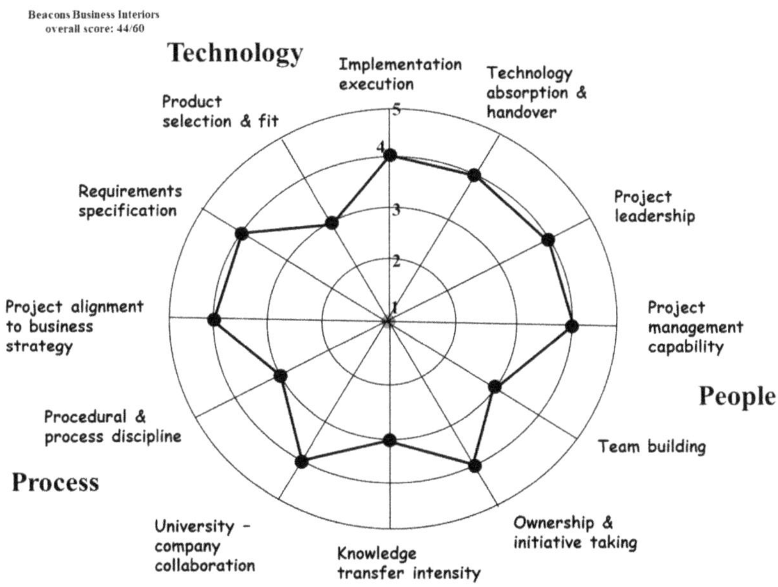

NEW E-TRADING SYSTEMS AT TPG DISABLEAIDS, HEREFORD

Background

As noted in the first case study in this chapter, TPG DisableAids is a family business founded in 1984, and is a provider of equipment for the elderly and disabled. Products range from daily-living aids, stair-lifts and mobility scooters, through to bespoke patient moving equipment installed into the fabric of homes and public sector buildings. Significant turnover comes from after-sales service. They have numerous suppliers and manufacturers, many of whom manufacture personalized products, such as stair lift products from Stannah. The company had an annual turnover of £3.9m in 2007-8, with stair lift products generating about one-third of turnover but over 50% of profits.

The company's market can be divided into different segments (National Health Service, local authorities, district councils, residential and nursing homes, private individuals). Nationwide, this is a multi-billion pound market, which is growing as the age profile of the population increases. At project outset in 2008, it was clear that the business opportunity was there to rapidly grow market share, particularly in the new market segments driven by public authority care management, insurance industry home equipment provision, and lifestyle products for the elderly. It was important that the company had the e-business trading capability to respond to the equipment and service requirements of the National Health Service (NHS) and related bodies at short notice as the elderly and disabled leave hospital and return to their homes. This alignment was seen as critical to the retention of larger public sector organizations and the company's expansion plans.

The company had trialed a CRM software package following the full review of information systems strategy discussed in the first case study in this chapter. The selected package was Le Grand CRM, which was used for storing marketing and customer data and was linked to the company's sales order processing software, although it was never used as an order entry system in the manner in which many CRM packages are. It was seen as a low cost pilot to help determine functional requirements, which were concluded within 9 months. The problems with interfacing this package with the company's existing legacy systems led the company to pursue a bespoke software development strategy to provide on-line order capture capabilities for its major customers. This was the focus of this second technology transfer project at the company.

Project Objectives

The project proposal (Technology Strategy Board, 2008) stated that "the NHS e-procurement initiatives require specific systems integration capabilities which the company do not currently have. The project will provide this capability and also act as a model for TPG DisableAids to drive efficiencies through its own supply chain via systems and process integration" (p.4). More specifically, the objectives were stated as follows:

- *"Research and develop a full top-level process map for the company's extended supply chain and identify opportunities for process integration.*
- *Identify technology issues, conflicts and requirements to exploit process integration opportunities. Specification of functional requirements and the technical architecture required to allow multi-point access to new TPG DisableAids systems and client systems via the web.*
- *Development and implementation of bespoke functionality as required to provide systems integration and web access.*
- *The driving through of process change to deliver cross-supply chain benefits.*
- *The embedding of new outward looking information culture and process change through intensive staff training and support" (Technology Strategy Board, 2008, p.5).*

Project Implementation

This was a 2 year project that started in late 2008, and the project was by and large implemented in line with the original objectives (Figure 7). The Associate, a project leader with appropriate technical skills, was engaged to develop and implement a web portal linked to the company's sales order processing legacy systems by bespoke middleware. From a technology perspective, the challenge was to upgrade the IT infrastructure to provide a solid platform to build an electronic trading capability through a web portal. This would be linked to the middleware infrastructure, thus filling the technology gap between the company's old legacy accounting system and new modern technologies possessed by key customers. The Associate, working with the IT manager, elected to use open source/freeware which provided a secure, reliable and flexible platform for development. Implementing freeware technologies in a live business environment required significant research and knowledge

Figure 7. The implementation plan for TPG DisableAids e-trading project (Robert James Howlett, Innovation through Knowledge Transfer. 2010, Springer –Verlag Berlin Heidelberg reproduced with permission of SNCSC)

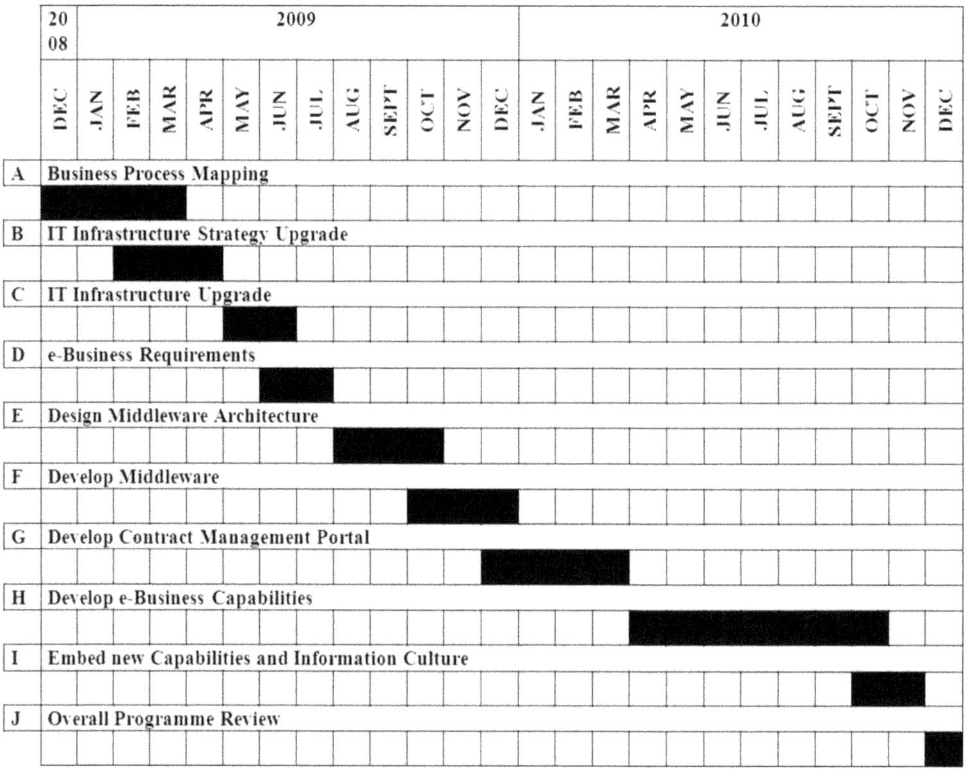

transfer. MySQL Community Server and PostFix Email Server - secure and reliable open source technologies – were both used in the project.

There was a range of technological challenges in developing and implementing the middleware, due to the technical and functional shortcomings of the legacy systems and their limited integration capabilities. After the middleware design stage, development comprised two main phases: first, the development and synchronization of a middleware database; and second, the synchronization of data in the data mart containing contracts information (Figure 8). Converting information from old file structures in the legacy accounting system to a modern relational database management system was problematic as there were only limited options available to convert the flat data files (DBF format) that existed in the legacy system (Sybiz Vision). Open source products were used to convert DBF files into modern SQL

Figure 8. TPG DisableAids systems architecture (Robert James Howlett, Innovation through Knowledge Transfer. 2010, Springer –Verlag Berlin Heidelberg reproduced with permission of SNCSC)

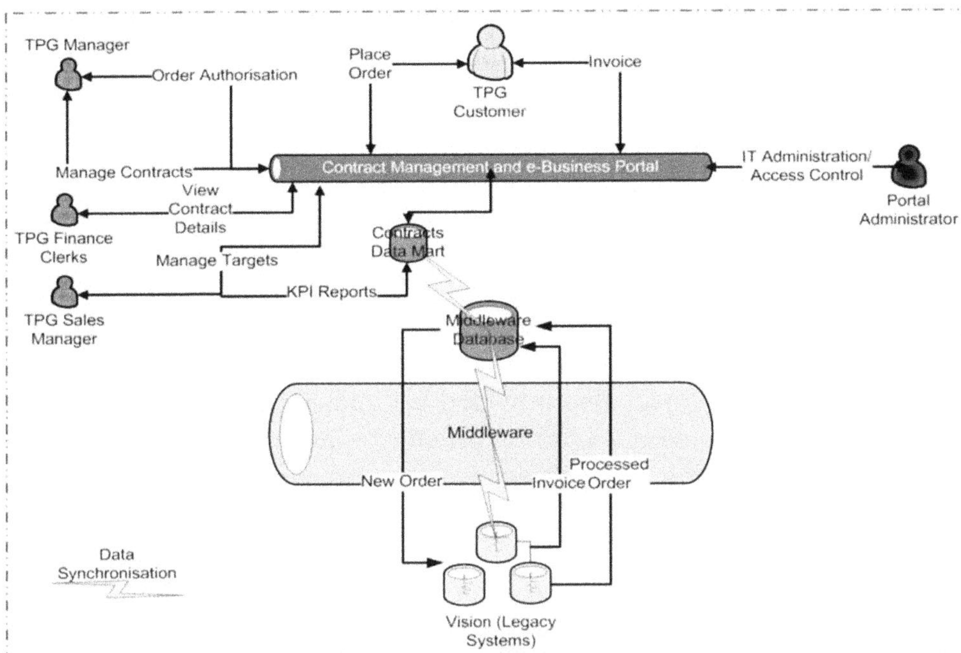

based information. Assuring quality and consistency of the data was another challenge. A well-designed architecture was key to accomplishing this critical phase of the project.

The development of the new portal for contracts management and e-trading was the critical deliverable from the project from a business perspective and was dependent on the stability and reliability of the middleware synchronization of the database, data mart and legacy systems. The portal architecture was based on the modern technology (Model View Controller/ J2EE) design pattern whereas the design of the middleware database was necessarily based on the old legacy accounting systems file structures. Extracting and transferring data to and from the legacy accounting system and the modern technology based portal was a collective effort, utilizing different open source products, freeware utilities and intelligent architecture design. This allowed the production of capabilities and functionality not possible in the old legacy systems.

Project Outcomes

The web portal went live in 2010 and has allowed the company to meet the e-trading demands of its key public sector customers, including new e-procurement/order capture capabilities to allow transaction processing with NHS and other key customers. The company had the new capability to trade electronically with local authority organizations responsible for the provision of disabled facilities grants and associated products and services, which could include trading via third party data transfer intermediaries or other similar agencies. The project also brought improved efficiencies throughout the order and sales processing procedures utilizing web portal technology, whereby order information is accepted over the web and returned to the customer as an invoice, thus minimizing the opportunity for human or machine error. The project also provided key management information for contracts to date, reduction in errors in pricing for larger customers and errors during invoicing, and for the creation of new contract pricing structures. The company's environmental impact was also improved by removing the need to print paper documents and post to customers. The company was able to maintain and grow its place in the market, turnover increasing to £4.5 million in 2014/15.

Analysis

Most aspects of this project were very positive (Figure 9) and the UK Government funding agency gave it a C rating. It was closely aligned to overall business strategy, it was well supported and lead by company management, there was significant technology transfer into the company, and there was a good "esprit de corps" within the company-university team. The one weakness that undermined the project somewhat was that the implementation, although technically sound, was not tied into on-going discussions and briefings with the company's key customers. This meant that the take-up of the system has been restricted and with hindsight, regular consultations with customers throughout the course of development and implementation would have been mutually beneficial. This critical activity was not carried out. The project thus delivered a technically sound bespoke product, but despite its potential, its use has been limited.

Nevertheless, as the Final Report (Technology Strategy Board, 2012) noted, the "project itself was to meet a direct requirement of our public sector

Figure 9. Change factors in the TPG DisableAids e-trading project

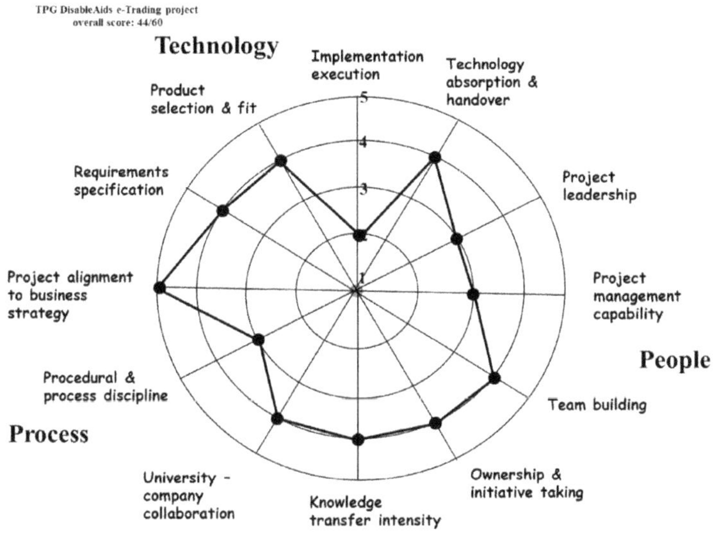

customers to secure their continued custom. This was met" (p.3); and the project did "allow development of future electronic trading and strategic decisions regarding our major customers" (p.3). It was the catalyst for significant process change within the company. Up to this point, processes and technology in the company were aligned to customer sectors on the one hand, and to suppliers on the other. The project required the company to develop cross sector processes for customer and supplier information management and inter-organization communication. Supply chain processes and operations have thus become aligned to both product and customer, rather than to solely customer sector or supplier. This differs from the customer-centric realignment that is often associated with new e-trading systems, and more closely represents the reality of operations in companies such as TPG DisableAids where they are aligned to both customer and to key suppliers from which they procure finished or near-finished products for assembly and sale.

CONCLUSION

The first two projects discussed in this chapter focused on process improvement, and the third (TPG DisableAids e-trading project) was mainly concerned with retaining and expanding the company's services to major clients, albeit allied to internal process improvement and the development of a new middleware software product. This latter project was the most radical of the three (Figure 10), but although it was successful from a technology perspective, it did not play the role envisaged in the original project plan in attracting new customers and growing market share.

These cases suggest a model for systems strategy development and implementation that is somewhat different from the idealized textbook models of the day, which often portrayed three tiers of information systems in a "software stack" of ERP system-data warehouse-business intelligence tools. For many SMEs and particularly SBEs, the required level of investment to achieve this was unrealistic, especially as most small companies had often acquired, over a number of years, individual software packages as point solutions. Faced with the requirement for a significant improvement in management information for both internal control and improved customer service, companies such as the two discussed in this chapter pursued a pragmatic approach that maximized the value of investments already made, and minimized the overheads involved in the purchase and implementation of additional point solutions and development of bespoke interfaces and reporting systems (Figure 11). In these companies, the innovative deployment of low cost solutions, allied to process improvement, was very effective in delivering enhanced analysis, management control and decision making in support of business growth: the technology transfer projects brought to these companies the required skills and technologies to develop appropriate strategies and then extract, interpret and match data from disparate systems to provide this key management information.

Figure 10. Product assembly companies: technology transfer focus and degree of innovation

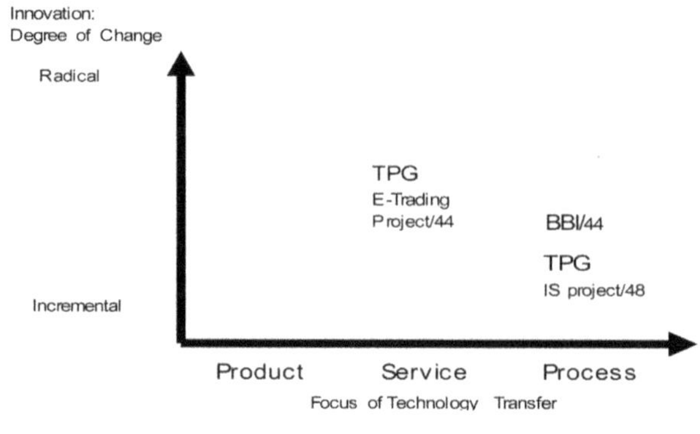

Figure 11. The idealised "software stack" (left) and technology options often used by SMEs (right)

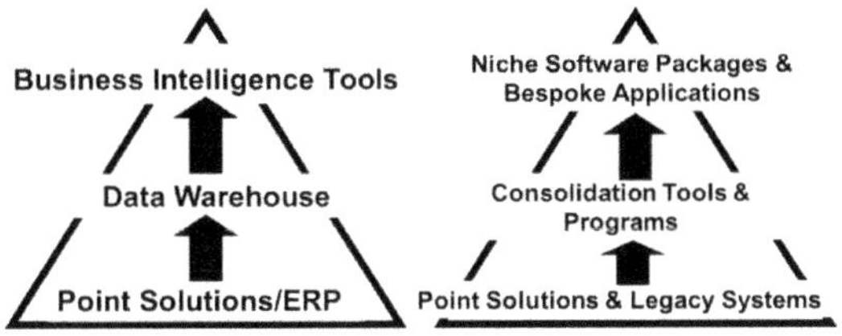

REFERENCES

Koopman, J. (1999, September). Effective alignment: strategy cannot succeed without it. *Canadian Manager*, 14-15.

Lee, H., & Ng, S. (1998). Preface to global technology and supply chain management. In H. Lee & S. Ng (Eds.), *Global technology and supply chain management* (pp. 1-3). Miami, FL: Productions and Operations Management Society.

Momenta. (2004a). *Knowledge Transfer Partnership Project Proposal and Application Form TPG DisableAids*. TTI Database Record, KTP Programme Office. (Unpublished)

Momenta. (2004b). *Knowledge Transfer Partnership Project Proposal and Application Form Beacons Business Interiors*. TTI Database Record, KTP Programme Office. (Unpublished)

Momenta. (2006). *Knowledge Transfer Partnership Final Report KTP 166 (Beacons Business Interiors)*. TTI Database Record, KTP Programme Office. (Unpublished)

Monk, C., & Wynn, M. (2007). *Knowledge Transfer Partnership Case Study 2 - Beacons Business Interiors*. Technology Strategy Board/University of Gloucestershire. Retrieved September 18th, 2018 from http://eprints.glos.ac.uk/

Stadtler, H., & Kilger, C. (2008). *Supply Chain Management and Advanced Planning – Concepts, Models, Software, Case Studies* (4th ed.). Springer. doi:10.1007/978-3-540-74512-9

Technology Strategy Board. (2007). *Knowledge Transfer Partnership Final Report KTP 692 (TPG DisableAids)*. TTI Database Record, KTP Programme Office. (Unpublished)

Technology Strategy Board. (2008). *Knowledge Transfer Partnership Project Proposal and Application Form TPG DisableAids*. Database Record. Technology Strategy Board. (Unpublished)

Technology Strategy Board. (2012). *Knowledge Transfer Partnership Final Report KTP 6964 (TPG DisableAids)*. Database Record. Technology Strategy Board. (Unpublished)

University of Gloucestershire. (2005). *Possible strategies for Bbi regarding their information systems*. (Unpublished)

Uppal, R., Wynn, M., & Turner, P. (2011). Providing e-business capability on a legacy systems platform: a case study from the KTP scheme. In R. J. Howlett (Ed.), *Innovation through Knowledge Transfer 2010*. Springer-Verlag. doi:10.1007/978-3-642-20508-8_18

Van Goor, A. R. (2001). *Demand and supply chain management: a logistical challenge*. Paper presented at 17th International Logistics Congress, Thessaloniki, Greece. Retrieved September 18th, 2018 from www.soleurope.net/servicesupport/documents/vangoor-salonique.pdf

Wynn, M., & Jones, P. (2017). Knowledge Transfer Partnerships and the entrepreneurial university. *Industry and Higher Education*, 1–12. doi:10.1177/0950422217705442

Wynn, M., Turner, P., Abbas, H., & Shen, R. (2009). Employing Knowledge Transfer to support IS implementation in SMEs. *Journal of Industry and Higher Education*, *23*(2), 111–125. doi:10.5367/000000009788146629

KEY TERMS AND DEFINITIONS

Data Base Files (DBF): Data base files store data and information, normally in tables, table fields, and field data values. The concept is normally linked to relational database management systems, which succeeded the old flat file databases in the 1980s.

Efficient Consumer Response (ECR): A strategy to improve the level of service offered to consumers by cross-supply chain cooperation by retailers, wholesalers, manufacturers and suppliers.

Information Systems (IS): Comprises software systems, applications, packaged software, software development tools, and operating systems. IS are one part of an overall Information Technology (IT) portfolio, that also includes hardware, networking, and telephony.

Key Performance Indicators (KPIs): Quantifiable measures used to assess the performance of an organization, a specific project, a department or even an individual. They often have financial and/or time related components.

Structured Query Language (SQL): The standard programming language for communicating with a database.

Vendor-Managed Inventory (VMI): A business model whereby the vendor of goods manages the supply of those products into the buyer's warehouse or store. This requires the buyer to provide details regarding inventory levels and order quantities to the supplier.

Chapter 6
Manufacturing Companies

ABSTRACT

One of the most debated areas regarding the introduction of new technologies into companies of all sizes is ERP systems implementation. These integrated software packages normally encompass the main transaction processing and information reporting requirements of a company, spanning sales order processing, financial management, human resource management, stock movement, and inventory control. It has led to a widespread debate in the literature regarding the respective merits of procuring and implementing an ERP system or deploying individual standalone software packages. The increased take-up of packaged software also coincided with the spread of business process re-engineering (BPR) to improve efficiencies and reduce overheads. The two became closely linked as BPR projects were frequently combined with the introduction of new software solutions. In this chapter, three such cases are reviewed, all involving major new packaged software implementations in manufacturing companies and all associated with varying degrees of process change.

BACKGROUND

The development of packaged business software to support all main business processes has been a major development in information systems over the past two decades. Packaged software for most mainstream business processes came to market in the 1990s as the spread of the UNIX operating system as a *de facto* standard for mini computers and the increasing dominance of the Intel

DOI: 10.4018/978-1-5225-7408-8.ch006

chipset led to a massive surge in the packaged software market. Building on the earlier MRP packages, other packaged software provided modules for sales order processing, ledgers, payroll and personnel as well as MRP, sometimes combined into one integrated package from one vendor – the ERP software suites of Oracle and SAP, for example.

Benson et al.'s (1992) "staple yourself to an order" example in the early 1990s illustrated well the potential for process improvement and introduced many practitioners to the concept of function vs. process. This was developed further in the 1990s by Michael Hammer and others (Hammer & Champy, 1993) who argued that process change was essential to business survival and acknowledged that this could be associated with new information systems. Hammer argued that there was no scientific definition of processes, but that there were three main business processes that could fit most company operations: "obtain a customer order", "fulfill the order" and "new product development".

These concepts were further developed by the emerging ERP software houses as they expanded the functionality of their products and started to differentiate between "process" manufacturing and "discrete" manufacturing, with different ERP products suiting their different requirements. Process manufacturers (for example, food and drink companies) would typically have short lead times, make to forecast and aim to hold minimum inventory. Discrete manufacturers (for example, the automobile companies and engineering companies) would have longer lead times, and often manufacture to fulfill specific orders. Chapter 1 provided two examples of technology transfer which centered on ERP products in contract packaging companies, these companies being akin to the process manufacturing model. In this chapter, the introduction of new systems in three manufacturing companies is discussed, two of which (at Dowty Propellers and Fixing Point) involve ERP products. Dowty Propellers, with its long lead times for specific orders for propeller blades, represents the discrete manufacturing model, whereas Fixing Point, a manufacturer of roofing and cladding components, aligns more with the process manufacturing model. In fact, Fixing Point elected to implement the same ERP product as that chosen by Brecon Pharmaceuticals in Chapter 1.

The third case study is of SKF AeroEngine Bearings, an engineering company in the aerospace sector. Whilst many such companies have deployed ERP products, the introduction of new systems and process thinking paralleled the development of the concept of full product life-cycle management for engineered products, from concept through design and engineering, to manufacturing, delivery and even product performance on client sites, spanning the new product development and order fulfilment processes identified by Hammer. This gave rise to a new set of information systems - Product Lifecycle Management (PLM) systems, which provide a framework for the monitoring and control of the product lifecycle in a particular business or product area. It is a compilation of business rules, methods, processes and guidelines as well as instructions on how to apply the rules in practice (Saaksvuori & Immonen, 2002). The PLM concept encompasses several systems. Ameri and Dutta (2005) described change management, document management, workflow management and project management as PLM systems that support concurrent engineering and streamlined product development processes. PLM seeks to fill the gap between enterprise business processes and product development processes.

The engineering industry also uses a range of other specialist (largely standalone) systems including Computer Aided Design (CAD), Computer Aided Manufacturing (CAM), and Computer Aided Engineering (CAE) packages, which are often interfaced with ERP, MRP and data warehouse/ business intelligence systems. Many of these types of systems feature in the technology transfer project at SKF AeroEngine Bearings discussed below.

ERP SYSTEMS AND PROCESS INTEGRATION AT DOWTY PROPELLERS, GLOUCESTER

Company Background

Dowty Propellers is a part of Smiths Aerospace. This was a division of Smiths Group when the project was undertaken, but is now part of General Electric. Smiths Aerospace is a leading global provider of innovative solutions to builders and operators of civil and military aircraft, from transport to fighters, from unmanned aerial vehicles to helicopters and regional jets. Smiths Aerospace then employed about 11,000 people in around 40 countries. During the 2005 financial year, it reported circa $2 billion revenues in Europe and North America.

The company was part of the mechanical systems business of Smiths Aerospace, with over 60 years' experience in the production of aircraft propeller assemblies. The company was recognized as the industry leader in the manufacturing of composite blades using resin transfer molding technology. At the start of the project in 2004, the company had 181 staff, and could thus be considered an SME, albeit being also part of a larger international group. There were two main divisions in the company - the manufacture of original equipment (OE) and the aftermarket business, which mainly concerned Dowty Propellers repair and overhaul (DPRO) operations, where the repair and warranty work and the overhaul of the propeller systems was undertaken. 70% of the business was related to OE and the remaining 30% came from DPRO.

An ERP system (Syteline) had already been installed in Dowty OE, but DPRO was functioning under a different ERP system (Fourth Shift). The company wished to integrate the two divisions from a process and systems perspective, via standardization on one ERP product. Combined with this, there was a broader initiative at Group level to migrate the company over a period of years to the SAP ERP product.

Group IS Strategy and Project Objectives

A key objective of Smiths Aerospace overall business strategy at the time was to strengthen their position as a "Tier 1" global supplier of integrated aircraft systems. Both Smiths Group as a whole and Dowty Propellers recognised the need to underpin their business strategy with improved systems and business processes. Smiths Aerospace was undergoing a major re-structuring with departmental functions such as human resources, finance and IT being centralised as "shared services". SAP had been chosen as the ERP solution that would ultimately support all of the Group's main functions and processes, although initially this was to be implemented to run the finance function only. The plan was to migrate all major businesses across the Smiths Aerospace division to the new "shared services" by 2008, and hence implement the SAP Financials module and integrate it with their existing business systems. This was to be the first step in the staged migration to the entire SAP suite. As regards the Dowty Propellers ERP project, as the Final Report (Momenta, 2006) noted, "this created an entirely new focus and plan for the project and also a new urgency, as the SAP implementation was very high priority and planned to tight timescales" (p.3).

This was undoubtedly going to be a difficult and lengthy process involving a number of interim solutions and short-term interface developments. As a result of company acquisition over many years, Smiths IT portfolio included

Figure 1. Main components of Smiths Group systems architecture 2006

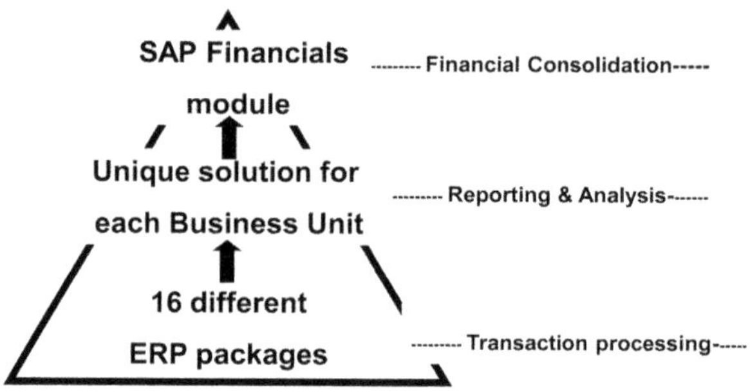

no less than 16 different ERP packages (Figure 1). A significant first step in rationalisation was to select just 4 of these 16 ERP packages that could co-exist alongside SAP, and would integrate, for the mid-term, with the SAP Financials module. These four packages are shown in Figure 2 and included, significantly, the Syteline ERP package, but not Fourth Shift.

The Syteline ERP had been implemented in the Dowty Propellers OE business in 1998, and was generally successful in supporting core business processes, albeit with significant package modification. DPRO had previously been located in a different part of the Smiths organisation, and had implemented the Fourth Shift ERP product. Dowty Propellers wanted to re-engineer its business processes around the two main value streams emanating from the OE manufacturing company and the DPRO "aftermarket" operations. A common IT platform was identified as a fundamental building block of the re-engineering process. In addition, Dowty Propellers had been identified within the Group for the roll out of the SAP Financials module. The business had to undergo extensive business process change to allow for a smooth implementation of Syteline and decommissioning of the Fourth Shift product,

Figure 2. Rationalisation of Smiths Group systems architecture 2006-8

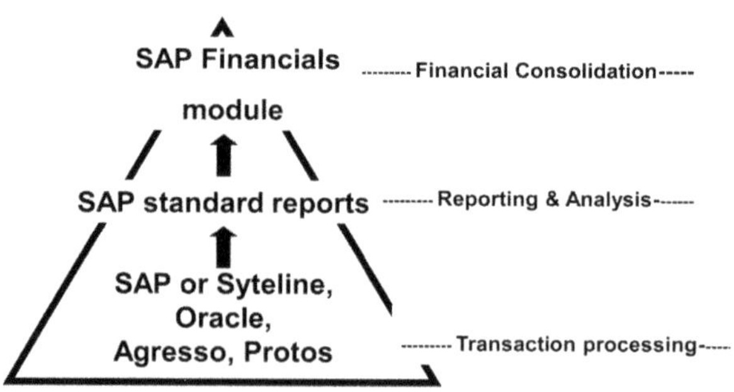

so that the SAP roll out could encompass both business units – Dowty OE and DPRO – with both businesses running on the Syteline system. The challenge was compounded by the fact that the version of Fourth Shift being run was very near to being "out of maintenance" (i.e. not supported by the software suppliers), adding more urgency to completion of the project. Process re-engineering, alongside the implementation of the Syteline ERP system at the DPRO business, were at the centre of the technology transfer project.

Project Implementation

Business analysis was carried out at an early stage of the project to identify the current status of the IT architecture at DPRO; this was necessary to determine and plan how best to link the two businesses together within the given short timescales. The business analysis revealed a system that had grown organically to support the needs of the business as it experienced a period of rapid growth. As Fourth Shift was unable to fulfil all business

Figure 3. End-user databases at DPRO in 2004 (indicated by the cylindrical symbol)

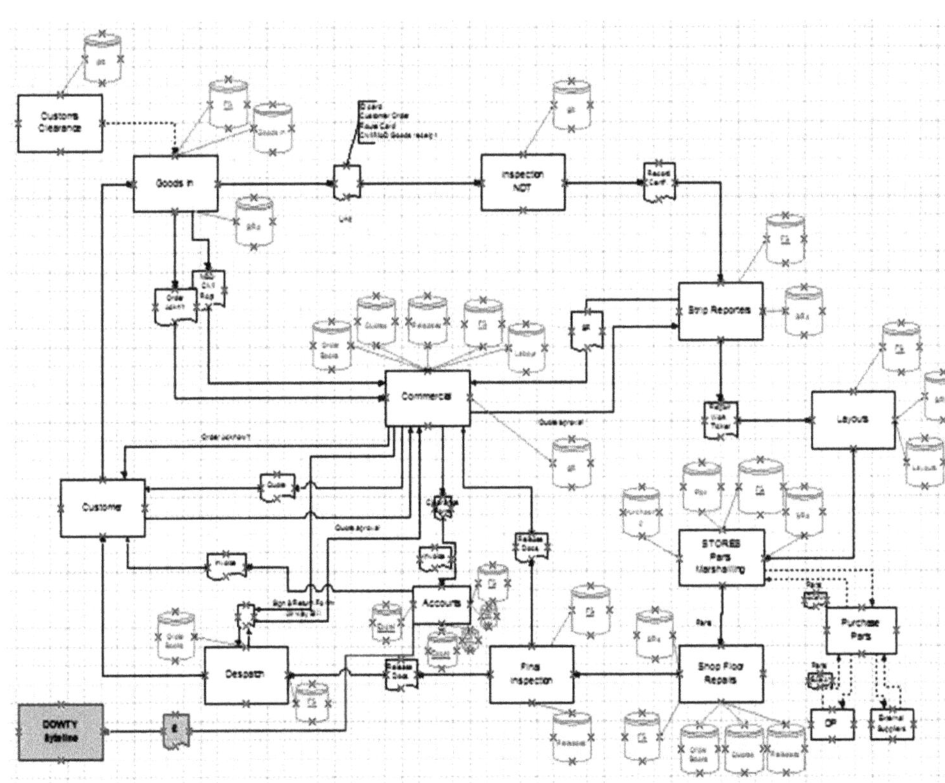

information requirements, enterprising computer end-users had developed over thirty standalone databases to help them run the business (Figure 3). Whilst this supported business operations reasonably well at the time, the company's management recognised the need to rationalise and simplify the company's information systems as the company entered a new phase of growth: the improved functionality of Syteline would reduce the reliance on end-user databases.

The proliferation of these databases had increased the support demands on the IT department, and the department had also lost much of its Fourth Shift expertise. The migration to Syteline therefore made support easier to deliver, and this was within the core IT skill set of Smiths Aerospace. It also facilitated change control and standardisation of the IT architecture. By adopting Syteline, the DPRO business was better aligned with Group IS strategy.

The project reviewed, improved and standardised core processes across the two businesses so that both companies could be supported by one configuration of Syteline. Gap analysis was carried out to identify the business requirements and map these to the functionality of the Syteline ERP system. Wherever possible, system changes were kept to a minimum, as Syteline had already been mapped to SAP in the OE business. A new IT architecture was devised and the number of peripheral databases was drastically reduced, leaving just four end-user Microsoft Access databases connected to the main ERP system. Communications between the two sites in Gloucester was also improved by linking the DPRO site to the ERP server hosted at the OE site. Systems piloting, user acceptance testing and training were progressed in accordance with the project plan.

The system was delivered as per plan and in time to allow work to continue integrating Syteline at the OE site with SAP. The new system went live at DPRO in March 2006, and no major problems were reported after go-live, only teething issues that were directly related to system configuration and user access rights. This was recognised as a potential issue on the plan as there was not enough time to cater for all the necessary system administration work before the go-live date. Business management agreed to compromise in some areas in order to have the DPRO business integrated into the OE business by the agreed date, but a post-implementation project phase was authorised to address these minor issues.

System Benefits

Systems integration across the two businesses helped Dowty Propellers to align IT with their business strategy. At the same time, the move to Syteline allowed compliance with Group policy of implementing SAP as the main financials package, without this representing a challenge to the two hitherto separate businesses, which could have involved sacrificing profitability, transparency and internal control.

An integrated systems strategy - instead of discrete, project-based activities managed as separate functions in departmental silos - enabled the company to achieve agility and improve revenues, enhancing productivity and ultimately optimising business performance. Integrating the two businesses on a unified systems platform allowed the organisation to reduce its IT complexity and obtain more business value from their IT investment, by allowing the DPRO business to be included in the SAP roll out.

In summary the main business benefits drawn from this systems and business integration project were:

- Processes were re-engineered and aligned to corporate business strategy; operations were streamlined and the use of corporate resources was optimised.
- Customer service was enhanced by providing one point of contact within Dowty Propellers.
- The complexity of the existing IT infrastructure and architecture was reduced by integrating the two business functions around one ERP system.
- Data and information were consolidated and standardised across the two businesses, facilitating improved access and reporting.
- It supported and progressed the Group information systems strategy, by preparing the businesses for an SAP implementation, and also ensured the flow down of benefits from the Group SAP project to the Dowty Propellers' business.
- IT efficiencies were improved by reducing the number of databases and systems requiring support.

Analysis

This project was given a B grade by Momenta, the UK Government funding body, and it generally delivered well against its stated objectives. There were nevertheless a number of difficulties that required particular focus to allow the project to progress. These included very tight timescales, as the project was on the critical path for the rollout of the SAP Financials module, and hence was constrained by the go-live date of the latter project. This did not allow enough time to complete a thorough re-engineering of business processes, and in addition the scope of the system implementation had to be restricted to the minimum to achieve the project end date. In these circumstances, project management capability and team building were of particular importance, and the project scored well on these factors; but implementation execution was nevertheless impaired by these constraints. Due to the tight time frame, there was very little contingency built into the programme, so any unexpected or unplanned for issue could have put the overall project at risk. The project go-live date was delayed on two occasions due to external factors, but the resultant small slippage had no significant knock-on impact on the SAP Financials project.

Figure 4. Change factors in the Dowty Propellers ERP project

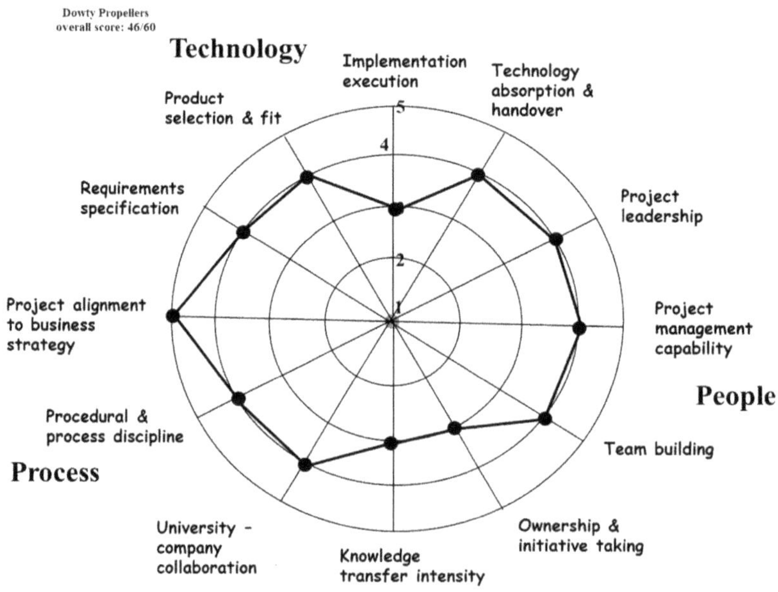

One major issue that put the project in jeopardy was the large number of parallel-running projects that were going on at Smiths, all competing for the same resources. The centralisation of both the finance and IT departments, the need for system changes to support a significant new customer of Smiths Aerospace, and the support of day-to-day business operations, all required certain key resources, which were all also involved in the DPRO systems project. Prioritisation proved to be a great challenge as the business had to constantly revise the allocation of resources and take decisions that allowed all projects to move forward at an acceptable pace. In these circumstances, the alignment of the project with overall business strategy was a key strength in arguing for resources, and the finance director noted that one major benefit of the liaison with the university was the availability of "a dedicated resource to manage key initiatives to drive the business forward" and that the Associate "played a significant role in systems delivery" (Monk & Wynn, 2008, p.1). Business leadership and commitment were also strong, with the finance director taking personal charge of the project, and chairing the project steering groups. This was particularly significant in resolving prioritisation and resourcing issues, and coordinating activities in the two business units.

This summary overview is reflected in the change factors diagram for the project (Figure 4). Project alignment to business strategy was crucial in underpinning the final successful implementation of the project, even if implementation execution was hampered by a number of business issues. The project scored well on all other factors. Although knowledge transfer intensity was only moderate, the university supervisor and Associate had wide experience of ERP projects and project management methodologies, and this contributed to the continuity of the project when problems were encountered and creative solutions needed to be found.

ERP IMPLEMENTATION AT FIXING POINT, CHELTENHAM

Company Profile

Fixing Point is a family business based in Cheltenham that designs, manufactures and distributes a wide range of non-standard, high quality, technically advanced products to the roofing, cladding and walling sectors of the construction industry. When the project started in 2006, the company had 53 staff and turnover that year was £5.4m.

Project Rationale

Fixing Point had implemented a number of standalone information systems using different technologies and running on disparate unconnected hardware platforms. The functionality of these systems was limited in a number of areas, notably in product costing, and management information was produced from a number of standalone packages and a range of spreadsheets, across four main sites around the UK. This was labour-intensive and produced data discrepancies, which impacted on customer service. Stock management across these sites was also problematic, and resultant "out-of-stocks" and inadequate resource capacity to meet deadlines resulted in unacceptable customer service levels.

A major project objective was the integration of order processing, manufacturing and stock control for the flashing and fabrication product group. In order to achieve this objective, Fixing Point elected to replace their legacy systems with one integrated package - an ERP system - and re-engineer business processes to support the company's growth plans. Fixing Point wanted to establish a common hardware platform and associated technical standards across all three product divisions, using standard procedures and practices which would aid the culture shift to a "one team" approach within the company. The new system needed to reduce the amount of duplicated processing and eliminate the need for spreadsheet based control systems. A significant improvement in the quality and availability of data was required.

Package Selection

The project commenced in July 2006 with an original planned duration of 18 months (Figure 5). The overarching objective was "to implement a new Enterprise Resource Planning system that will support the current and future information requirements of the company and its main business partners" (Momenta, 2005, p.2). Acquiring an ERP package that matched user requirements was considered critical to achieving project success, and after discussion with colleagues from the university, the Project Board (comprising senior management from the company and three university staff) agreed to use a combination of standard methodologies to guide the package selection process and related project management issues. A simplified version of the PRISM Buy-Build methodology (Figure 6) was used in conjunction with some elements of the PRINCE2 project management methodology. First,

Figure 5. The initial 18 month project plan for the Fixing Point ERP implementation (Momenta, 2005, p. 38)

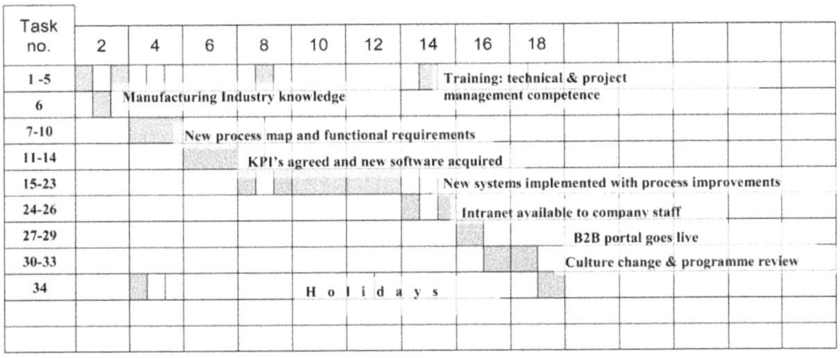

Figure 6. Elements of the PRISM Buy-Build Methodology used at Fixing Point (Wynn et al., 2009)

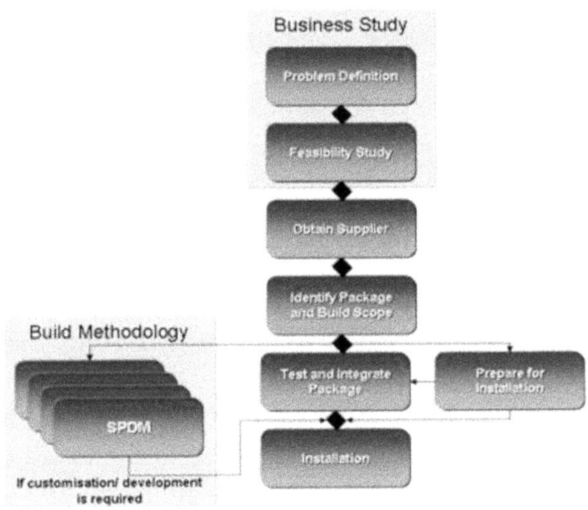

business processes were mapped at a high level and key users were interviewed to establish what systems currently existed, how they were performing, and current and future information needs. This allowed the project team to identify issues and information gaps in the company. Questionnaires were used to establish where and how key data items were treated – particularly product and customer data. Of the current systems, where were these key data items entered, processed and reported upon? Were there problems with data inconsistencies, and why? The findings were used as the basis for the production of key areas for improvement, KPIs and the list of user requirements for circulation to potential suppliers. Suitable ERP vendors for a manufacturing SME were selected and invited to send proposals. These activities took about 3 months in total.

An initial short-list of preferred vendor proposals was drawn-up from which four vendors were selected for a systems demonstration; from here, the preferred final two suppliers were requested to make further presentations. Structured workshops and detailed discussion on functionality, user requirements and price negotiation followed. This process took a further 5 months. At the end of this phase, Fixing Point chose the EFACS E/8 ERP system from Exel Computer Systems Ltd. This is a component based ERP package that allows some customisation and flexibility in the way it is implemented - the package can be adapted to fit specific functional requirements, which was considered of significance given the company's specific requirements for product costing.

Project Implementation

Training was arranged for the main users on the key functional aspects of the new system, and the Associate took responsibility for mapping the new system's capabilities against Fixing Point's business processes. Activities during this period included unit testing of the main systems modules, migration of key data to the new system, customisation of the system where necessary, and a range of additional workshops for training, enrolment and decision making. All these activities were overseen by the Project Board. The implementation phase was completed within 6 months, albeit with some delays incurred by competing business priorities and unexpected events. For example, changes in staffing meant that systems users were not able to complete their testing within the planned time frame. The Project Board elected not to rush with the implementation, but rather to ensure a high quality implementation. Systems testing was done thoroughly, and adequate training for all users was provided to underpin a smooth transition to the new software and a successful embedding of new processes across the company. The system went fully live after a month of parallel running in July 2008, this approach ensuring that users had ample time to familiarise themselves with the new system and associated procedures. The lengthy package selection process had a knock-on effect of extending the project end date by 6 months. This was agreed by the university in conjunction with the Government funding body and the company themselves.

Project Benefits

The EFACS ERP software suite effectively supported the Fixing Point business operations at the time, which saw a 10% growth in sales year on year, with turnover growing to over £8.0m in 2009/10, and a steady continued increase in profits and margins. The new system allowed the integration of data and information processing in the main business functions of the company – materials flashing and fabrication, pricing, sales, manufacturing and financial management. Data entry was streamlined, eliminating costly duplication; and stock management was done more quickly and more accurately, minimising over-stocking and under-stocking problems across the company's four UK sites. The training programme was continued and systems users in key departments were given the skills to become expert in report generation and business intelligence gathering from the ERP database. Key users were also trained to become systems administrators to maintain and manage the system.

Analysis

The project at Fixing Point delivered a major systems implementation, which is not an easy task. As Koch (1999) humorously remarked "the inherent difficulties of implementing something as complex as an ERP is like teaching an elephant to do the hootchy-kootchy" (p.4). It was this achievement that warranted the B grade given to the project by the Technology Strategy Board, and the project management and technical capability of the Associate was significant in achieving this success. This is indicated in Figure 7, which is taken from the company finance director's presentation at one of the Project Board meetings. Nevertheless, there were a number of internal and external factors that impacted project timescales, resulting in a 6 months slippage. The Final Report noted:

"The problems encountered during the project and their solutions were as follows:

- *Selection of the supplier. Fixing Point did not want to choose in haste as they were making a 10 year investment. Thus, a more careful approach was taken where all decision makers had to be comfortable with the selection of the supplier.*

Figure 7. Project support benefits to Fixing Point (University of Gloucestershire, 2007)

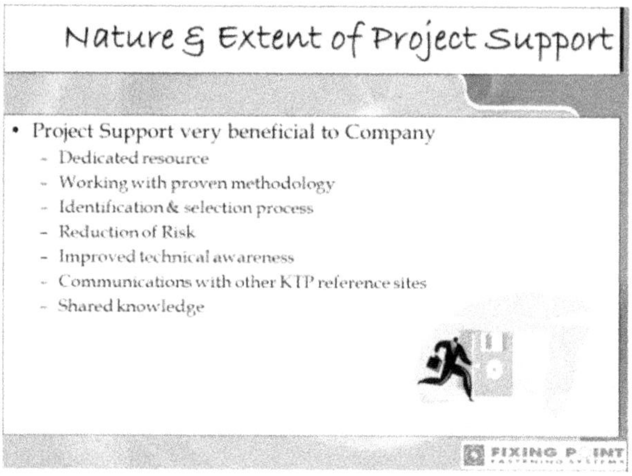

- *Restructuring of chosen supplier. The unexpected circumstances of the chosen supplier had pushed the initial target of implementation from March 2007 to June 2007.*
- *Flooding. In July 2007, Cheltenham, Gloucester and Tewksbury suffered their worst flooding ever following severe heavy rain lasting several days. The training and user workshops that had been scheduled for July with external consultants were postponed until September.*
- *Staff changes. Some key users were identified to lead testing, but were unable to do so because of unexpected staff departures. Thus, the schedule was revised to accommodate the new staffing situation.*
- *Inertia. There was some reluctance to switch from the old, familiar system to the new, unfamiliar one. Thus a special effort was made to visit all of the branches to introduce and explain the new system" (Technology Strategy Board, 2008, p.2).*

The change factors diagram (Figure 8) indicates the relative strengths and weaknesses of the project. Most factors scored well, but the technical and project management aspects were strongest, underpinning an effective specification of requirements and software package selection. The durability of the Associate and her project management knowledge and experience were important in getting the project "over the line" when a combination of internal and external factors threatened project delivery. The alignment of the project with the company's business strategy meant that there was no turning back once underway, and sponsorship of the project by, and general

Figure 8. Change factors in the Fixing Point ERP project

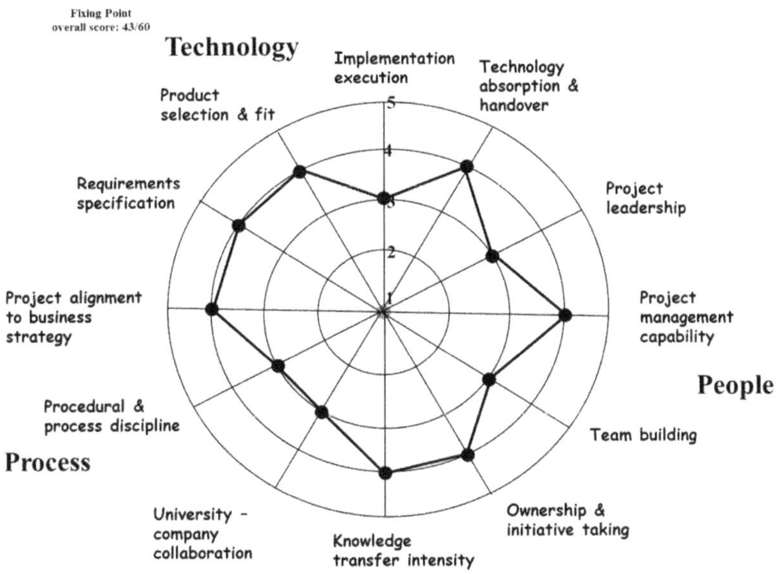

support of, the finance director were equally important in providing continuity and additional funding when required. Nevertheless, decisions made by the company's senior management in the broader interests of other company priorities impacted negatively on project implementation, team building and procedural and process discipline. Knowledge transfer intensity was high and technology absorption and handover were well executed with extensive training and sound product documentation. With hindsight, the diligent and thorough specification of requirements, which resulted in the selection of a package well able to support the specific needs of the company (notably in supporting a complex product pricing process) was of particular importance in the eventual success of the project.

ENGINEERING SYSTEMS INTEGRATION AT SKF AEROENGINE BEARINGS, STONEHOUSE, GLOUCESTER

Company Background

When SKF AeroEngine Bearings UK (SKF) embarked on this technology project in 2007, it employed 250 staff and had an annual turnover of circa

£17m. The company, which was part of the SKF Group, had 7% of the world market in aero engine bearings, the majority of which derived from major aircraft engine manufacturers. The project proposal (Technology Strategy Board, 2007) noted that "to retain existing market share and create opportunities for new business, the company needs to be more efficient in providing engineering data and more cost effective to be price competitive" (p.2). A major upgrade of their shop floor engineering systems and associated technology infrastructure was seen as essential if the company was going to be more effective in a highly competitive market. The company were also the first in the Group to implement such an upgrade, and thus product selection and implementation on the shop floor would set a precedent for other Group companies.

The project involved the upgrade and integration of a range of systems specifically geared to shop floor manufacturing operations, notably programmable logical controllers (PLCs) and associated supervisory control and data acquisition (SCADA) systems and computer aided design (CAD) tools. The project also encompassed Product Lifecycle Management (PLM) systems, which had evolved from earlier Product Data Management (PDM) systems. A PDM system can be seen as a subset of a PLM system. In the 1980s, engineers had recognized the need to manage the increasing volumes of design data produced by, and contained in, a range of automated and semi-automated systems, such as CAD files, specification and requirement documents, and CAM and CAE analytics. PLM and PDM systems allowed the user to store and control all product data, manage document issue levels, maintain Bill of Materials and immediately visualize the relationship between parts and assemblies.

Project Objectives

The overarching aim of the project – named "Engineering Systems Integration" (ESI) - was to develop and implement integrated engineering information systems to support future business competitiveness based on improved customer responsiveness. The implementation of a new PLM system as well as new CAD/CAM packages, both linked with ERP and shop floor data systems, was seen as essential to support the company in delivering big contracts to its main customers. The main software products that were acquired were NX (CAD/CAM) and Teamcenter (PLM), which were to be implemented and integrated with other systems over a two year period. More specifically, the objectives of the ESI project were to:

- Create a business process map which encompassed the company's main business processes
- Implement NX (CAD/CAM) and manage the NX project pilot
- Manage the basic Teamcenter (PLM) implementation
- Evaluate and cost justify the need for, and benefits of, other software packages
- Define and implement the integration of NX and Teamcenter with CMMs and CNCs
- Implement managerial reports using business intelligence tools (Cognos Pro-IV)
- Implement a shop floor planning system

Project Implementation

The project comprised 8 stages (Figure 9) and elements of the PRINCE2 project management methodology were used to run and control the project (Wynn, Shen, & Brandao, 2008). At the first meeting of the Project Board, it was recognised that for the project to be successful, there would need to be three parallel running and inter-related streams of work – Process, Technology and People/Training streams. Each of these streams was allocated a senior manager to act as overall owner and coordinator. The initial stage focused on industry knowledge gathering, and process mapping. The Associate held a series of workshops with management and shop floor staff to get agreement

Figure 9. Project timetable for ESI project at SKF (Robert James Howlett, Innovation Through Knowledge Transfer Smart Innovations, Systems and Technologies, 2010 Springer –Verlag Berlin Heidelberg reproduced with permission of SNCSC)

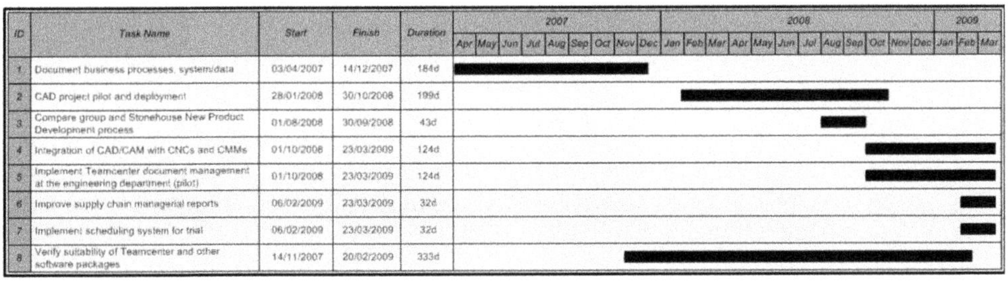

to the process maps and start to identify areas for process improvement. Some Six Sigma techniques (such as spaghetti diagram and lead time analysis) were applied in order to understand where priority focus was required. Other sites (and even competitors) were visited and analyzed, to ascertain what technologies were used to address each of the company's issues, thereby establishing the Associate's credibility and role brief with the company's engineers.

In stage 2, NX - the new CAD system - was implemented on the shop floor, moving users to a three-dimensional CAD environment for the first time. This reinforced project momentum and broadened support for the project as a whole. As part of stage 3, the process maps (Figure 10) were revisited, and the project team spent some time examining the overall group processes for product development and manufacturing, and the implications of Group software strategy (which favored the NX/Team Centre PLM software combination). Stage 4 then refocused on software implementation and developed links between the CAD/CAM functions and the shop floor machines (CNCs and CMMs), where the major focus was on product quality and precision after each iteration of the product's development. The initial raw material underwent a sequence of operations that involved CNCs and/or manual machines shaping the component, with CMMs verifying whether those machines achieved the tolerance required by each stage on the product line. To have a machine program downloaded to the CNC, a Direct Numerical Control (DNC) system was used, providing program version control and the logging of program interruptions and necessary amendments. SFK were the first aerospace business within the corporate Group to achieve the integration of the 3D CAD/CAM system with the CNC and CMM machines.

Figure 10. Top-level business processes at SKF

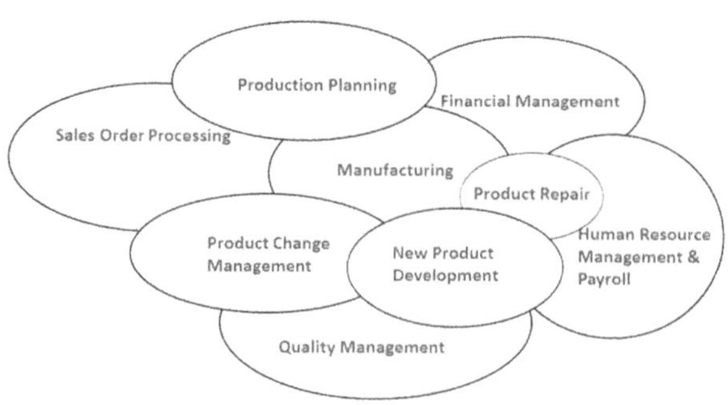

The Teamcenter document management system was installed in the engineering department in stage 5, delivering significant time saving benefits for information retrieval. Until then, the company was still using print copies as their masters for auditing and even for product changes. Through use of the digital copy, with all product information also being held in the PLM system, the management and maintenance of product development and change control became faster and more secure. Stage 6 reflected a change of focus as business priorities and staff personnel changed. The Associate was tasked with assessing possible information improvement across the broader supply chain, and his main achievement at this stage was the development of new management information, using the Cognos Pro-IV business intelligence tool running on the CHESS MRP system.

As regards process change, customer facing product channels were introduced to drive new efficiencies on the shop floor. This divided the product portfolio into categories, and machines were aligned in sequence on the shop floor creating product category lines. Before the introduction of this concept, the shop floor was laid out with the same type of machines being clustered together in cells. The process change team were charged with implementing this element of the overall project, but key information was needed to make the most of the concept. Before the new channels were fully implemented, it was necessary to identify "bottleneck machines" in each channel, so that appropriate redesign of the channel machine configurations could be made. The Cognos business intelligence tool was used to generate the daily reports required by the channel implementation team.

The supply chain function was further developed in stage 7, where the Associate undertook package evaluation for other supply chain processes, notably scheduling, leading to the selection and pilot implementation of a new scheduling software package. Prior to this, lead-time calculations were difficult to estimate due to the number of orders, the quantity of operations for product development and the limitations of the corporate MRP package in visualizing the impact of any changes on the shop floor. Due to a long lead-time for aerospace product development, some customers were providing forecast orders for the following 3 years and, to be able to secure these orders, the company had to prove it had available capacity over that period. The new scheduling system provided the ability to simulate those conditions and report an accurate picture of future capacity. This stage completed a major overhaul of the company's IS portfolio (Figure 11).

Figure 11. Selected software solutions and systems architecture at SKF. Software packages implemented at SKF were CHESS (MRP), Teamcentre (PLM), NX (CAD/CAM) Seiki (Advanced Scheduling) and Cognos Pro-IV (Business Intelligence/Reporting). (Robert James Howlett, Innovation Through Knowledge Transfer Smart Innovations, Systems and Technologies, Springer –Verlag Berlin Heidelberg reproduced with permission of SNCSC)

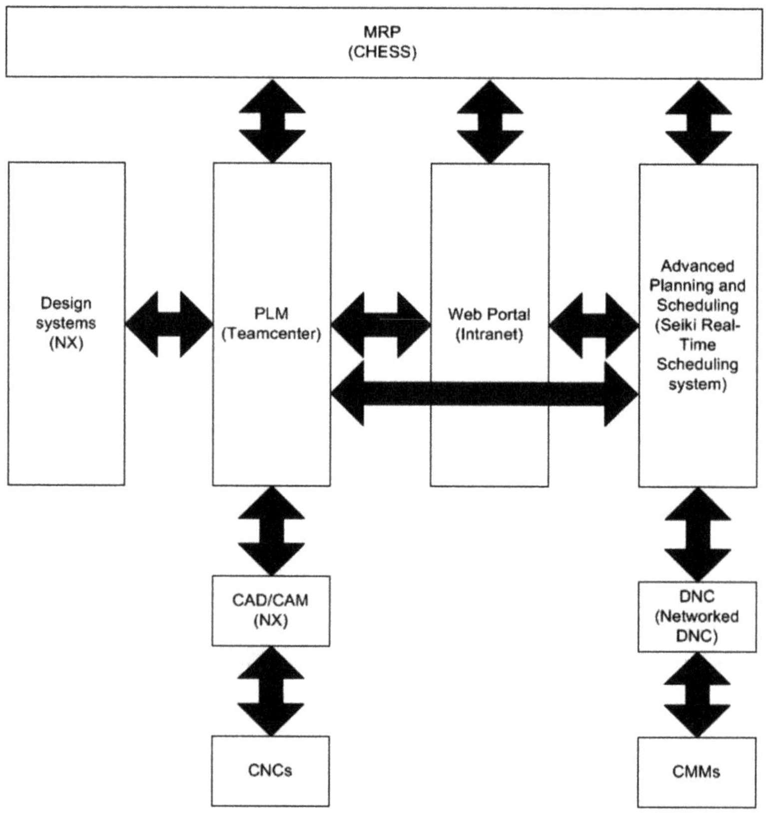

Project Benefits

In terms of technology transfer, new knowledge was generated, transferred and acquired in the process mapping, software package evaluation and implementation processes. The creation and validation of process maps fostered a better understanding of the relationships between departments, which was used in a customer audit to demonstrate new development flows. The re-organization of the design manuals in the design portal improved document management and provided a tool for document retrieval. "The integration of the CAD\CAM system with the CNC and CMM machines was the first site of the aerospace group division to achieve this integration and provided technical knowledge of machine programs implementation that can be used elsewhere in the group" (Technology Strategy Board, 2009, p.4).

Hitherto, the company had very little in-house IS resource, but the project helped develop new staff capabilities to further embed the new CAD and Teamcentre systems; and the comparative analysis of company business processes with those at Group level facilitated alignment of processes and software deployment across the group. The new PLM system provided document management, 3D visualization, and process management functions. Document management was used across the business while the 3D visualization was mainly used in the engineering design department and on the shop floor.

The implementation of this new systems portfolio was accompanied by a reorganization of the shop floor to halve product lead times for bearings from 90 days to 45 days, with a forward target of 30 days. This involved a radical change in shop floor structure from an organization that reflected management

Figure 12. Shop floor process alignment around channels at SKF (Robert James Howlett, Innovation Through Knowledge Transfer Smart Innovations, Systems and Technologies, Springer, 2010–Verlag Berlin Heidelberg reproduced with permission of SNCSC)

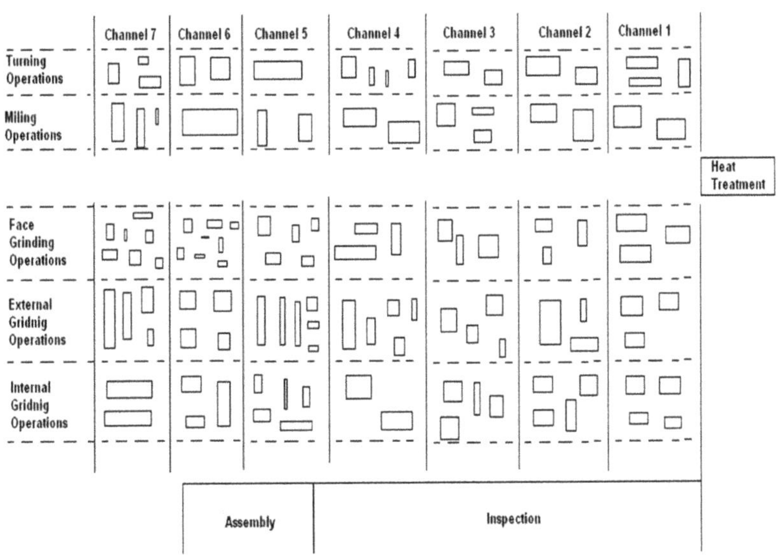

functions to one based around product-customer channels, in which 80% of operations required for product development and delivery were contained within each channel. Seven main channels were identified that crossed and utilized elements of the old functions (e.g. heat treatment, grinding, final view-assembly, inspection - Figure 12). This resulted in faster product flow and less work in progress, and eliminated product data on systems not linked to specific orders.

Savings achieved by implementing the CAD/CAM and PLM systems were predicted to payback the investment cost within three years and exceed a £1m net benefit within 10 years. The implementation of the new scheduling and capacity planning system allowed the company to take new contracts worth £1.5m per year in 2010, 2011 and 2012. The working relationship and availability of information on the company's engineering database between departments was substantially improved as a result of the project.

Analysis

This project introduced new systems and re-engineered processes in a complex shop floor engineering environment and also delivered new and improved

Figure 13. Change factors in the engineering systems integration project at SKF

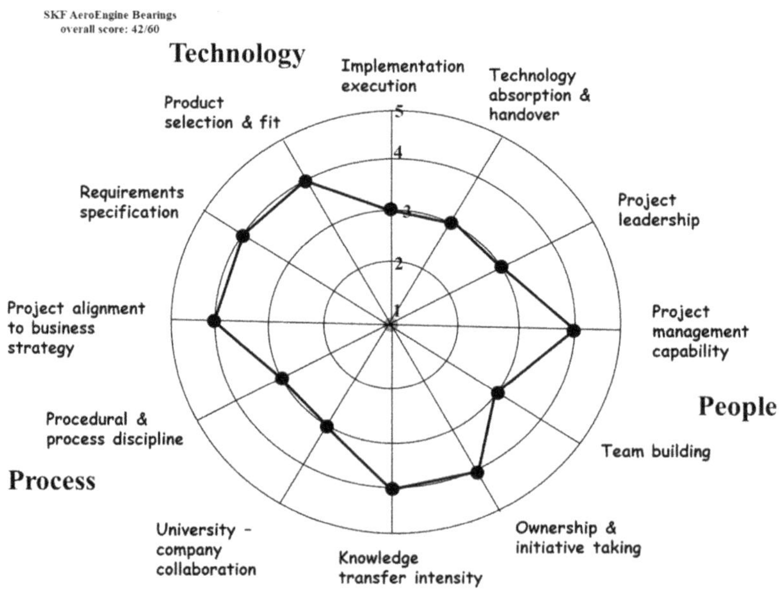

management information. It was rated as a B by the Technology Strategy Board, and the change factors diagram (Figure 13) highlights the relative strengths and weaknesses of the project. There were several staff changes in company management directly involved in the project across the two year period, and this had a negative impact on project momentum and direction. Team building and procedural discipline suffered somewhat, and the project changed focus several times. However, due to the flexibility and all round ability of the Associate, knowledge transfer intensity was high and new software products were appropriately selected and well implemented; and overall the project was closely aligned to the company's strategic direction, which ensured continued support despite the managerial changes.

The project involved change in processes and procedures, and training was necessary on a number of implementation fronts: design/development, programming, new product development, and change management. Significant effort was dedicated to helping engineers to come to terms with, and embrace, the new processes and systems, which in the end produced the institutionalization and embedding of new ways of working. Not surprisingly, there were difficulties in implementing some of the new technologies. For example the CAD/CAM system was geared more to automotive businesses, and some difficult design shapes needed in the aerospace industry had to be converted from the old 2D system to the new 3D CAD/CAM system. Another example was the PLM system and its data classification solution. The level of detail handled by the classification was limited in comparison with the levels of detail needed by an aerospace business, and systems customizations had to be made to attend to this requirement. Implementation execution proved difficult, but ownership and initiative taking was strong overall, notably by the Associate, and this helped ensure overall success of this a multi-faceted project.

In summary, the project exhibited a high degree of technology expertise that supported the selection, implementation and integration of new engineering systems, allied to significant process re-engineering on the shop floor. However, the changes in senior management staff affected project direction and overall continuity to some extent, but the alignment of the project to overall business strategy and the project management capability of the Associate and his business colleagues ensured a successful project outcome.

Figure 14. Manufacturing companies: technology transfer focus and degree of innovation

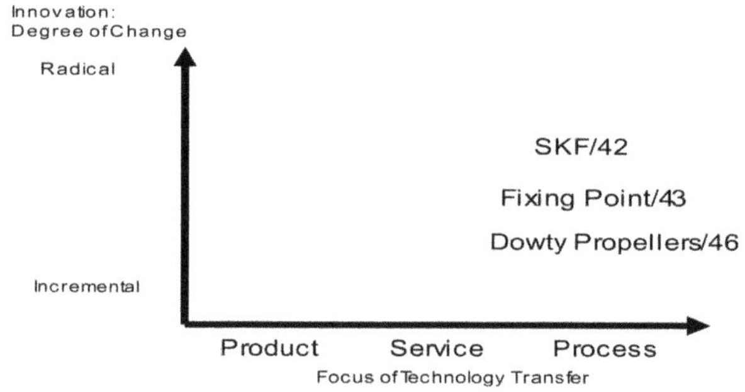

CONCLUSION

The three projects at these manufacturing companies all involved major systems implementations allied to significant process change. None of the projects was especially radical, although the project at SKF was perhaps more so given the degree of change involved (Figure 14). In these circumstances, the role of the Associate was critical, being a dedicated knowledgeable resource to instigate, promote and manage key initiatives to drive the technology transfer forward.

The right balance needs to be struck between delivering new technology and undertaking the associated process change. McAfee (2003), with reference to major software package implementations, noted that "the coordination, managerial oversight and marshaling of resources needed to implement these systems make for a change effort like no other" (p.83). Process re-engineering can help large projects such as those discussed in this chapter to deliver major benefits in terms of efficiency gains and customer service, but there is a danger that it can delay implementation and stall the project. At Dowty Propellers, the need to go live with Syteline to meet the SAP Financials module roll-out meant that process re-engineering was curtailed, and at SKF there were a number of such delays, in part because of the extensive training needed to support staff in new ways of working; and at Fixing Point, it was the package selection process, in particular, that pushed back the project end-date by 6 months. These cases evidence the need to build in contingencies, in terms of both timescales and required resources, in projects of this nature and scale.

REFERENCES

Ameri, F., & Dutta, D. (2005). Product lifecycle management: Closing the knowledge loops. *Computer-Aided Design and Applications*, *2*(5), 577–590. doi:10.1080/16864360.2005.10738322

Benson, P., & Shapiro, V., Rangan, K., & Sviokla, J. (1992). Staple Yourself to an Order. *Harvard Business Review*, *70*(4), 113–122.

Hammer, M., & Champy, J. (1993). *Reengineering the Corporation*. New York: Harper Collins.

Koch, C. (1999, December 22). The ABCs of ERP. *CIO Magazine*. Retrieved September 24th, 2018 from http://wikifab.dimf.etsii.upm.es/wikifab/images/d/da/The_ABCs_of_ERP.pdf

McAfee, A. (2003). When too much IT knowledge is a dangerous thing. *MIT Sloan Management Review*, *44*(2), 83–89.

Momenta. (2005). *Knowledge Transfer Partnership Project Proposal and Application Form Fixing Point*. TTI Database Record, KTP Programme Office. (Unpublished)

Momenta. (2006). *Knowledge Transfer Partnership Final Report KTP 4454 (Dowty Propellers)*. TTI Database Record, KTP Programme Office. (Unpublished)

Monk, C., & Wynn, M. (2008). *Knowledge Transfer Partnership Case Study 3 - Dowty Propellers*. Technology Strategy Board/University of Gloucestershire. Retrieved September 19th, 2018 from http://eprints.glos.ac.uk/

Saaksvuori, A., & Immonen, A. (2002). *Product Lifecycle Management*. New York: Springer.

Technology Strategy Board. (2007). *Knowledge Transfer Partnership Project Proposal and Application Form SKF AeroEngine Bearings*. Database Record. Technology Strategy Board. (Unpublished)

Technology Strategy Board. (2008). *Knowledge Transfer Partnership Final Report KTP 1092 (Fixing Point)*. Database Record. Technology Strategy Board. (Unpublished)

Technology Strategy Board. (2009). *Knowledge Transfer Partnership Final Report KTP 1120 (SKF AeroEngine Bearings).* Database Record. Technology Strategy Board. (Unpublished)

University of Gloucestershire (2007, June 6). *Project Board/Local Management Committee 3 presentation.* (Unpublished)

Wynn, M., & Brandao, R. (2010). Using information systems to drive process change: an aerospace industry example from the Knowledge Transfer Partnership scheme. In R. J. Howlett (Ed.), *Innovation Through Knowledge Transfer Smart Innovations, Systems and Technologies* (Vol. 5, Part 2, pp. 125–134). Springer-Verlag. doi:10.1007/978-3-642-14594-0_13

Wynn, M., Shen, R., & Brandao, R. (2008). *Managing IS projects in SMEs – Tailoring the PRINCE2 methodology.* Paper presented at Conference on Money, Economy and Management, World Business Institute, Imperial College London, London, UK. Retrieved September 16th, 2018 from http://eprints.glos.ac.uk/

Wynn, M., Turner, P., Abbas, H., & Shen, R. (2009). Employing Knowledge Transfer to support IS implementation in SMEs. *Journal of Industry and Higher Education, 23*(2), 111–125. doi:10.5367/000000009788146629

KEY TERMS AND DEFINITIONS

Business Process Re-Engineering (BPR): An approach to change management pioneered in the 1980s and 1990s that often went hand-in-hand with major information systems implementations. The tasks required to obtain a specific business outcomes were often radically redesigned, and could result in speedier response times, less manpower, and lower inventories.

Computer Numerical Controls (CNCs): A method for automating the control of machine tools (drills, boring tools, lathes) through software embedded in a microcomputer attached to the tool.

Computer-Aided Design (CAD) System: A software system used by engineers (and other professions) to create precision drawings or technical illustrations. The output can include both two-dimensional (2-D) drawings but also three-dimensional (3-D) models.

Computer-Aided Engineering (CAE): The deployment of a range of computer software to aid in engineering analysis. This may include the use of computer software to simulate performance to improve product designs, and the optimization of products, processes, and manufacturing tools.

Computer-Aided Manufacturing (CAM): The use of software to control machine tools in shop floor manufacturing. It uses computer software and machinery to facilitate and automate manufacturing processes. The concept overlaps with CAE and is sometimes seen as its successor. It is often used in conjunction with computer-aided design.

Computerized Maintenance Management System (CMMS): Computer software designed to simplify maintenance management operations. It schedules, tracks and monitors maintenance activities to provide a range of reports and information on cost, component items, and personnel activities.

Direct Numerical Control (DNC): A general term used in shop floor operations to mean the networking of CNC machine tools.

Product Data Management (PDM) System: Used to manage product data and process-related information. It may include data generated by computer-aided design (CAD) systems. A PDM system may be part of a Product Lifecycle Management (PLM) system, which has a wider range of functions.

Product Lifecycle Management (PLM): A systematic approach to managing the various steps in the lifecycle of a product, from its design and development to its sale and indeed its ultimate disposal. A PLM system can integrate data relating to the product, relevant human resources and business processes. It may integrate with a company-wide Enterprise Resource Planning system.

Programmable Logical Controllers (PLCs): Ruggedized shop floor computers used for industrial automation. They can be deployed in the control, monitoring and automation of specific machines or production lines. Data gathered in PLCs is often uploaded to SCADA systems.

Supervisory Control and Data Acquisition (SCADA) Systems: Data collection and reporting systems that provide information on shop floor production and packaging processes. They are usually linked by network to monitoring devices like PLCs and other electrical controls.

Chapter 7
Conclusion

ABSTRACT

This chapter attempts to pull together the lessons learnt from the 19 cases reviewed in this book and to provide some guidance for industry professionals and academics undertaking technology transfer projects in the future.

INTRODUCTION

This book examines how technology transfer has operated in university-industry projects undertaken in SMEs via the KTP scheme. This comes at a time when the Confederation of British Industry (CBI) in the UK have recently called for the Government to redirect some of the money it will get back from the European Regional Development Fund after the country's exit from the European Union to support businesses and encourage them to implement better technologies. The report (Confederation of British Industry, 2017) concludes that "too few businesses take up the technologies and management practices employed by those leading the way" (p.4), and the group wants to establish a list of key technologies that are needed for firms to become more productive. The report also suggests that extra research to find the best way to implement these technology projects could be funded through the Government's new industrial strategy (Wallace, 2017). This aligns with recent research by Zhang (2018) who studied knowledge exchange in 133 UK universities and reported that "the results suggest that the knowledge exchange policy should not be implemented alone; rather, it should be developed in conjunction with

DOI: 10.4018/978-1-5225-7408-8.ch007

industrial and innovation policies" (p.80). This chapter therefore sets out some of the key lessons learnt from the technology transfer projects studied in the book, which are summarized in Table 1.

TECHNOLOGY TRANSFER CHANGE FACTORS

A review of the change factors in the nineteen projects studied allows some comparisons to be made. The technology, people and process change factors were assessed fairly similarly on average (3.8/3.9/3.7 respectively out of a maximum of 5 - Table 2). However, within these three change dimensions, there were some significant variations, and certain of these factors stand out

Table 1. Change factors summary

Company Name	Code	Industry Sector	Main Technology Focus	Change Factors/60	Funding Body Rating/ A-E
Allpay.net	a	Service Industry	Product	50	B
AuraQ	b	Software Company	Product	36	D
Beacons Business Interiors	c	Product Design & Assembly	Process	44	B
Beaumont Travel	d	Software Company	Product	51	B
Brecon Pharmaceuticals 1	e	Contract Packaging	Process	47	B
Brecon Pharmaceuticals 2	f	Contract Packaging	Service	51	B
Building Solutions	g	House Builder	Process	46	C
C&G Services	h	Service Industry	Process	46	C
Contrapak	i	Contract Packaging	Process	38	N/A
Dowty Propellers	j	Manufacturing	Process	46	B
E-Business Services	k	Software Company	Product	43	C
Energist UK	l	Service Industry	Process	42	D
Fixing Point	m	Manufacturing	Process	43	B
Muddy Boots	n	Software Company	Service	46	C
Optimum Consultancy	o	Service Industry	Process	54	B
Pegasus Retirement Homes	p	House Builder	Process	51	B
SKF AeroEngine Bearings	q	Manufacturing	Process	42	B
TPG DisableAids 1	r	Product Design & Assembly	Process	48	B
TPG DisableAids 2	s	Product Design & Assembly	Service	44	C

Table 2. Change factors assessment by project

Project	a	b	c	d	e	f	g	h	i	j	k	l	m	n	o	p	q	r	s	Av
Technology																				
Requirements specification effectiveness	5	3	4	4	3	5	4	5	4	4	4	5	4	4	4	5	4	4	4	4.2
Product selection & fit	4	4	3	4	3	5	4	4	4	4	5	4	4	4	4	4	4	3	4	3.9
Implementation execution	4	3	4	4	4	4	4	3	2	3	4	4	3	4	4	4	3	4	2	3.4
Technology absorption & handover	5	2	4	4	4	4	4	3	2	4	3	3	4	4	5	5	3	5	4	3.8
																			Technology Average	**3.8**
People																				
Project leadership	4	4	4	5	5	5	5	4	3	4	4	4	3	4	5	4	3	4	3	4.1
Project management capability	5	4	4	4	4	4	4	4	3	4	4	4	4	4	5	5	4	4	3	4.1
Team building effectiveness	3	2	3	4	4	3	3	4	3	4	2	3	3	3	4	3	3	4	4	3.3
Ownership and initiative taking	4	3	4	5	4	4	4	4	3	3	3	3	4	4	5	5	4	5	4	4.2
																			People Average	**3.9**
Process																				
Strategy alignment	3	3	4	5	5	5	5	4	5	5	5	5	4	5	5	5	4	4	5	4.5
Procedural and process discipline	5	3	3	4	4	4	3	3	3	4	3	4	3	4	4	4	3	3	3	3.5
University-company collaboration	4	2	4	4	4	4	3	4	3	4	3	3	3	3	5	3	3	4	4	3.5
KT intensity	4	3	3	4	3	4	3	4	3	3	3	2	4	3	4	4	4	4	4	3.5
																			Process Average	**3.7**
Score Totals	50	36	44	51	47	51	46	46	38	46	43	42	43	46	54	51	42	48	44	
Funding Body Grade (A-E)	B	D	B	B	B	B	C	C	N/A	B	C	D	B	C	B	B	B	B	C	

COMPANY/PROJECT KEY: a: Allpay.net; b: AuraQ; c: BBI; d: Beaumont Travel; e: Brecon Pharmaceuticals 1; f: Brecon Pharmaceuticals 2; g: Building Solutions; h: C&G Services; i: Contrapak; j: Dowty Propellers; k: E-Business Services; l: Energist UK; m: Fixing Point; n: Muddy Boots Software; o: Optimum Consultancy; p: Pegasus Retirement Homes; q: SKF AeroEngine Bearings; r: TPG DisableAids 1; s: TPG DisableAids 2.

as being of particular importance in determining project outcomes. Project alignment to overall business strategy was strong in nearly all projects, and this emerges as perhaps the key success factor. Even when other aspects of the project were not so successful – for example at E-Business Services and Energist UK - the fact that these projects were seen as critical to projected company growth ensured continued management support and eventually resulted in successful project outcomes. On the other hand, at Allpay.net, the lack of clear alignment of the original project proposal with company strategy led to it being abandoned early in the project cycle and replaced by a significantly different development project; and at AuraQ, a lack of consistency in business strategy in a difficult trading environment meant that the project contributed very little to the company's subsequent development.

The two weakest factors overall were team building effectiveness and implementation execution. Some Associates experienced difficulty in adapting to the company culture and disciplines, and in some instances they left the projects before the project end date. Fortunately, this was not of any great significance in any of the projects, but it did hamper team building, as did the high turnover of internal staff in one or two of the projects. The technology factor averages suggest a pattern that is reflected in many projects. The initial requirements specification is generally conducted to a high standard (average of 4.2), the product fit and selection is then completed reasonably well (3.9), but then the need to address and resolve a range of unforeseen technical, human and resourcing issues impacts implementation execution (3.4), causing project overruns in time and budget. Many projects experienced this "mid-project dip" (Fixing Point, SKF AeroEngine Bearings, TPG DisableAids 2, C&G Services, Contrapak), although in most cases these problems were finally overcome to allow project completion and technology handover (3.8).

The projects at Optimum Consultancy, Pegasus, Beaumont Travel and Brecon Pharmaceuticals (Lab project) achieved the highest change factors scores and were all assessed as "very good" (B grade) by the UK Government funding body. Optimum was very strong on project related skills and capabilities, and Brecon's precise requirements specification and product selection for the laboratory equipment reflected the experience and knowledge of both the Associate and the company line manager. Beaumont Travel was probably the most radical of all the projects. It changed the core business of the company, but despite its immediate impact (and thus very positive assessment), the company's initial success was not maintained in the mid-term. The project at Pegasus was strong in most regards, but notably in the leadership and commitment provided by the in-house management. A general conclusion is that all factors in the three dimensions of change need to be constantly addressed and kept in balance as technology transfer projects progress through their lifecycle.

TECHNOLOGY TRANSFER PROJECT FOCUS

The majority of the projects were concerned with using new systems and other technologies to improve internal (but often customer facing) processes (Figure 1). Some of these projects achieved significant process re-engineering in parallel with implementing these new technologies. At Pegasus Retirement Homes, for example, the implementation of a new CRM system represented a step change in the company's ability to analyze marketing and sales data and hence increase sales. Their newly installed VPN and wireless technology allowed sales staff to work remotely from different Pegasus offices and access the new CRM system. As a result Pegasus required fewer administrative staff to support an expanding sales team and the project dramatically increased the rate of sale through enhanced CRM capability and improved communications with staff and customers. Similarly, at C&G Services, a new web-based course booking and administration system (Course Booker) provided faster information processing and better access to key data for both customers and

Figure 1. Impact and focus in the nineteen technology transfer projects
COMPANY/PROJECT KEY: a: Allpay.net; b: AuraQ; c: BBI; d: Beaumont Travel; e: Brecon Pharmaceuticals 1; f: Brecon Pharmaceuticals 2; g: Building Solutions; h: C&G Services; i: Contrapak; j: Dowty Propellers; k: E-Business Services; l: Energist UK; m: Fixing Point; n: Muddy Boots Software; o: Optimum Consultancy; p: Pegasus Retirement Homes; q: SKF AeroEngine Bearings; r: TPG DisableAids 1; s: TPG DisableAids 2.

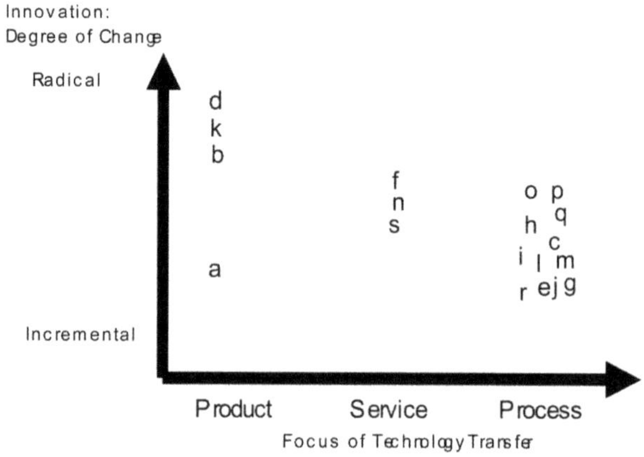

staff, allowing more time for proactive work with current and prospective clients. Clients' access to training course schedules and the associated booking procedures was significantly improved. Greater information visibility allowed staff more control over their workload, seen by clients in terms of improved and more proactive support facilities. At SKF AeroEngine Bearings, the implementation of a range of new systems was the catalyst for a re-engineering of shop floor processes to halve product lead times, resulting in faster product flow and less work-in-progress. These three companies, along with Optimum Consultancy, mentioned above, are the best examples of processes being radically changed - if not totally transformed - by the introduction of new technology.

Three companies attempted to introduce new technologies for service innovation. The project at Brecon Pharmaceuticals was arguably the most successful, where the introduction of an in-house laboratory services function had a major impact in terms of new customer attraction and bottom line revenue delivery. The company has grown significantly since project completion, having been acquired by the American international pharmaceutical services company AmerisourceBergen, to act as its hub for European operations, and now part of the multi-national PCI group. At TPG DisableAids, a provider of equipment for the elderly and disabled, the second project entailed the design, development and implementation of a range of bespoke middleware and data warehouse products to connect its existing legacy systems to a web-front end, providing the systems capability to respond to the equipment and service requirements of the NHS and related bodies at short notice. At Muddy Boots, the improved deployment of the Microsoft CRM software system - and associated new training modules, documentation and process improvement - helped the company make a step change in its global service provision.

Four of the projects had the objective of product innovation through technology change, and this proved the most difficult to achieve in the mid-term. At Allpay.net, the project was initially set up to develop and market a new software product for the management of social housing. However, a strategic review at company level led to the abandonment of the project early on. A totally revamped project, which focused on developing middleware products to link existing in-house legacy systems to improve their integration, went ahead, and was successful in its new objectives, which aimed to improve company data transfer and reporting processes. The project at AuraQ aimed at product innovation but failed to deliver against targeted objectives for different reasons. The company's business plan at that time entailed a move towards software solutions rather than services, and specifically to provide bespoke software components based on the Metastorm BPM product range. However, shortly after completion of the project, Metastorm were bought out by OpenText who changed development plans for the product. The envisaged expansion based on the new Metastorm templates did not take place. At Beaumont Travel, mentioned above, early success in transforming the company to a specialist software house was not maintained in the mid-term.

In summary, the focus and impact of innovation in these projects varied. Most (twelve of the nineteen studied) centered on achieving process improvement, with four projects attempting product innovation and three trying to innovate services to external customers. Using technology for process innovation is likely to be most successful, and the majority of these projects achieved significant improvement in bottom-line performance following the introduction of more effective information systems and associated process change. Using technology for product innovation is likely to encounter a number of problems that process and service innovation usually do not – notably changes in the external environment that affect the market potential of new products, and the consequent reassessment of risk factors that may lead to delays or reduction in internal project support.

ENTREPRENEURIAL THINKING - THE CATALYST FOR TECHNOLOGY TRANSFER

The term entrepreneurship is widely used in business, government and media circles, and has overlaps with the innovation concept, and is thus of relevance in understanding how technology transfer can most effectively be achieved. In particular, entrepreneurial activity often provides the spark that initiates technology transfer, and as such it is of value to consider the contextual factors that engendered this process in the projects studied here. Kao (2006) suggested that "entrepreneurship is the process of doing something new and something different" (p. 69), whilst Bruyant and Julien (2000) have similarly argued that "entrepreneurship is concerned first and foremost with a process of change, emergence and creation" (p.173). Carlsson et al. (2013) see entrepreneurs as "individuals....acting independently or within organizations, to perceive and create new opportunities" (p.914).

Wynn and Jones (2018) examined the contextual factors in evidence in four technology transfer projects and concluded that entrepreneurial activity played a key role in developing and implementing these projects in small companies. These findings (Figure 2) are reinforced by a wider review of the nineteen projects featured in this book. Regarding the broader socio-economic environment, the influence and outreach of the local university played the pivotal role in the instigation of these projects, and in the fostering of entrepreneurial activity within them. The University of Gloucestershire's strategic support for enterprise was important in providing the environment in which entrepreneurial activity was encouraged. There was, for example a commitment to "developing enterprising behaviour and a spirit of entrepreneurship", and "to provide a coherent and well integrated programme of support for businesses and employees" (University of Gloucestershire, 2012, p.12).

Once the projects were underway, the Associates also played an important role in the entrepreneurial development and management of the technology transfer at the operational level. In the Beaumont Travel project, for example, the Associate acted as team leader, investigated new market opportunities and added new functionality to the proposed new software, which in turn, met changing customer requirements. More generally, the academics acted as a conduit between the skills and expertise available from the university and the practical solutions needed by the company.

Figure 2. Contextual factors supporting entrepreneurial activity and technology transfer

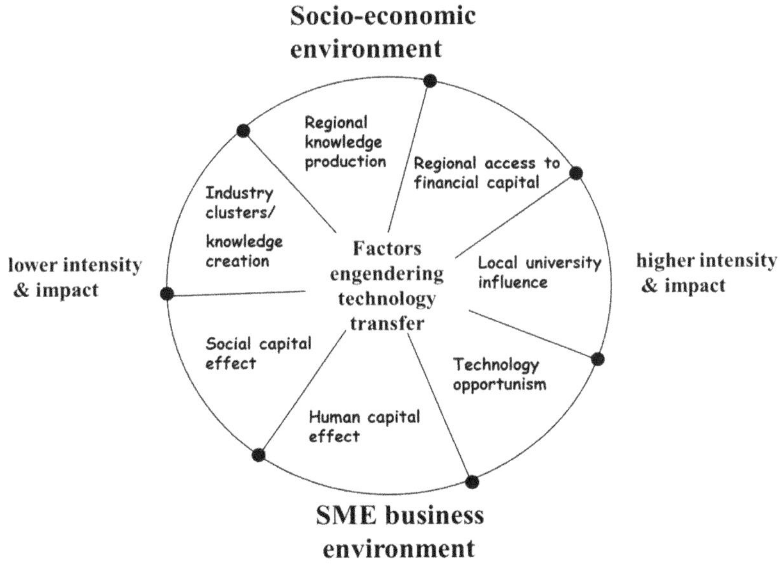

What Stam and Bosma (2015) identified as "regional knowledge production" and "regional access to financial capital" (p.329) were also important contextual factors in supporting entrepreneurial activity in many of these projects. The university held a number of events in the period 2003-2012, at which technology transfer opportunities were explained and follow-up meetings with interested companies were arranged. There were also a number of conferences including the annual "Growing Gloucestershire" conference and breakfast briefings to local companies in other locations such as Hereford. A range of seminars and briefings involving university staff and local industries spread knowledge and awareness about the KTP scheme in general, but also clarified the financial benefits of undertaking technology transfer projects within the operational and financial framework of the scheme. The access to financial capital was a critical factor in encouraging these companies to make a significant investment of their own time and human resources, as well as committing their own finances to these projects.

At the same time, the cumulative acquisition and application of skills and expertise from one project to others over time was a factor in generating new ideas and momentum for technology transfer. This aligns with the "knowledge creation and the presence of industry clusters" that Carlsson et al. (2013, p.915) identified as key influencing factors in developing entrepreneurial activity. For example, in 2010, six projects involving the University of Gloucestershire were underway with SMEs in the region (Figure 3), and this helped facilitate the cross-fertilisation of ideas and exchange of experience which promoted the development of new thinking. The university also held a

Figure 3. Technology transfer projects with locally based SMEs in 2010

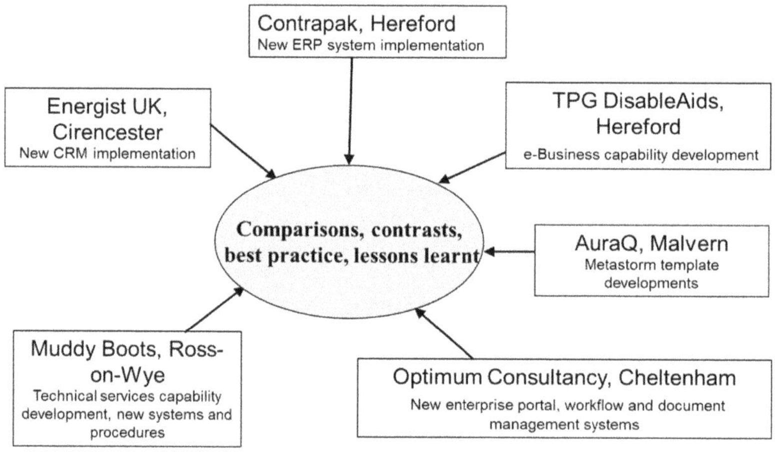

number of seminars for Associates working in parallel on the different projects to encourage exchange of experiences and application of methodologies and new ideas in their projects if appropriate.

There were also other factors within the business environment of the companies in which these projects were undertaken that encouraged entrepreneurial thinking in the context of technology transfer. A close knit family environment appears to have been a factor in allowing entrepreneurial activity to flourish in some of these projects – notably at TPG DisableAids, Beaumont Travel, C&G Services, Fixing Point and Energist UK. This points to "entrepreneurship as a social (family) phenomenon" (Stam & Bosma, p.329) being of some relevance, representing what they called the "human capital effect", with family members learning new entrepreneurial skills. In other projects, for example, at Beaumont Travel, there was also evidence of "a social capital effect" (p.329), whereby the owner's son was actively involved in the software development process, working alongside the Associate and the university academic supervisor. This resonates with the findings of de Wit-de Vries et al. (2018) who studied university-industry technology transfer and reported that "social capital is built through frequent face-to-face communication and workshops that facilitate interaction. This kind of communication improves insight into the partner's goals" (p.12).

But perhaps the critical link between technology transfer and entrepreneurism in these projects was what can be called "technology opportunism" - that is, the collective understanding between the university academics and company management of how the introduction of new technology could support the company's growth in a cost-effective manner through the KTP scheme. In these projects, this often focused on the introduction of new software packages or the development of software products either for sale or in-house deployment. Web-enablement of these systems and products was also a significant step forward for many of these companies. Looking to the future, however, the overall technology landscape has evolved somewhat, and the challenge now is to apply this opportunism and entrepreneurial thinking to a broader set of technologies.

THE CHALLENGE AND THE OPPORTUNITY

There were three main types of technology deployed in all but one of the projects discussed in this book. Four projects introduced new ERP systems; eight companies adopted a "best of breed" IS strategy in which they implemented point solutions, notably new CRM systems; and six projects centered on bespoke developments, either for in-house use or for the development of new company products (Table 3). Only at Brecon Pharmaceuticals, where the new laboratory equipment was installed, was the project of a different nature. The change factor scores are not significantly different between these three types of project, but the introduction of new ERP systems was probably the most difficult to implement, reflecting the scale and broad cross-company impact of such projects. Nevertheless, three of these four projects succeeded in their objectives and illustrate that change on this scale can be achieved with sound management and appropriate resourcing.

The challenge now is to extend and adapt this type of technology transfer project to a wider set of industries and technologies. In 2017, the CBI concluded that many UK businesses suffer from a "failure to adopt" the mainstream technologies currently available. The CBI Director General noted that "while the eyes of the business world can often be on the 'next big thing' in cutting-edge technology, too many firms are missing out on what's right under their nose" (Confederation of British Industry, 2017, p.27). They suggest that technology projects that firms should embark upon to become more productive include using e-purchasing technologies and implementing CRM (customer relationship management) systems, but also adopting cloud computing and enhancing cyber security (Wallace, 2017). More pointedly, IBM note that "cloud computing, mobility, social business, big data and analytics, and IT security technologies are evolving very rapidly", and that "as these technologies mature and converge, they are demanding a total re-examination of the underlying enterprise infrastructure" (IBM, 2015, p.2).

The challenges ahead are manifold. For the UK Government, developing and evolving frameworks like the KTP scheme to support businesses in adopting the mainstream technologies of the day is critical for future growth of smaller businesses and UK productivity overall. The UK Government's Industrial Strategy White Paper (H. M. Govmt, 2017) identifies a number of "grand challenges" that "are an invitation to business, academia and civil society to work together to innovate and develop new technologies and industries in areas of strategic importance to our country". These challenges include "putting the UK at the forefront of the artificial intelligence and data revolution", "being a world leader in shaping the future of mobility" and "harnessing the power of innovation to help meet the needs of an ageing society" (p.23). However, in parallel with these important strategic initiatives, there must also be continued and expanded support for the type of project and company discussed in this book, where the implementation of proven technologies involving mainly incremental change (rather than radical) is what is needed to grow and develop these businesses. As the CBI note "the UK has a great record and a strong reputation for cutting-edge innovation. .... But this is only part of the battle. Once these innovations are created, they are not being taken up as quickly as they should be across the economy. UK firms must be proud to be 'Magpies', picking up the innovations that other businesses have tested and proven" (Confederation of British Industry, 2017, p.14). This line of thought has been reiterated in other forums recently. For example, Overli (2018), in his recent article suggesting that "every business must be a technology business", notes that "the challenge is not to develop

Table 3. Technology transfer projects: technology implemented and project focus

Company	Enterprise Resource Planning System	CRM/Best of Breed Point Solutions	Bespoke Developments	Major Infrastructure Upgrade/Other	Project Focus
Allpay.net			YES		Product
AuraQ			YES		Product
Beacons Business Interiors			YES		Process
Beaumont Travel			YES		Product
Brecon Pharmaceuticals (1)	YES				Process
Brecon Pharmaceuticals (2)				YES	Service
Building Solutions		YES			Process
C&G Services		YES			Process
Contrapak	YES				Process
Dowty Propellers	YES				Process
E-Business Services			YES		Product
EnergistUK		YES			Process
Fixing Point	YES				Process
Muddy Boots		YES			Service
Optimum Consultancy		YES			Process
Pegasus Retirement Homes		YES		YES	Process
SKF AeroEngine Bearings		YES			Process
TPG DisableAids (1)		YES			Process
TPG DisableAids (2)			YES		Service
Project Change Factor Average	43	47	45	51	

the technology – there will be technology available to achieve your objectives. Rather, businesses need to reassess their goals – to work out what technology can help them achieve" (para.12).

The university sector has a vital role to play in meeting this challenge. Following the issue of the Industrial Strategy White Paper noted above, the UK Government appointed the Higher Education Funding Council for England to work with universities, research councils, Innovate UK and other knowledge exchange professional bodies to develop a new knowledge exchange framework. The framework "is intended to increase efficiency and effectiveness in the use of public funding for knowledge exchange (KE)" and to "further a culture of continuous improvement in university KE" (Higher Education Funding Council for England, 2017, para.2). It will be published later in 2018 and will comprise two main elements: on the one hand, there will be a set of principles for good practice, and on the other a set of related metrics for assessing knowledge exchange.

In this context, this book has set out a number of case studies that can be used as examples of good (and not so good) practice, and also a set of key change factors for technology transfer projects. There is an opportunity for academics and professionals to use this material in support of similar projects. These may now focus on a wider set of technologies, but the basic parameters for achieving successful technology transfer will remain the same. "Availability of knowledge and capital is key, driven by external collaboration and effective internal allocation of capital. The characteristics also reflect the culture, skills and capabilities within firms that support the successful take-up of best practice technologies and management" (Confederation of British Industry, 2017 p.20). It is hoped that the material and analysis contained in this book will assist those involved in such projects to better achieve successful outcomes.

REFERENCES

Bruyat, C., & Julien, P.-A. (2000). Defining the Field of Research in Entrepreneurship. *Journal of Business Venturing*, *16*(2), 165–180. doi:10.1016/S0883-9026(99)00043-9

Carlsson, B., Braunerhjelm, P., McKelvey, M., Olofsson, C., Persson, C. L., & Ylinenpää, H. (2013). The Evolving domain of Entrepreneurship Research. *Small Business Economics*, *41*(4), 913–930. doi:10.100711187-013-9503-y

Chesbrough, H. (2003). *Open Innovation: The New Imperative for Creating and Profiting from Technology*. Harvard Business School Press.

Confederation of British Industry. (2017). *From Ostrich to Magpie: Increasing Business Take-Up of Proven Ideas and Technologies.* Retrieved September 20th, 2018 from http://www.cbi.org.uk/index.cfm/_api/render/file/?method=inline&fileID=F840035B-0678-4A72-BAF8B7D170F47A07

de Wit-de Vries, E., Dolfsma, W. A., van der Windt, H. J., & Gerkema, M. P. (2018). Knowledge transfer in university--industry research partnerships: A review. *The Journal of Technology Transfer*. doi:10.100710961-018-9660-x

Govmt, H. M. (2017). *Industrial Strategy: Building a Britain fit for the future*. Retrieved September 20th, 2018 from https://www.gov.uk/government/publications/industrial-strategy-building-a-britain-fit-for-the-future

Higher Education Funding Council for England. (2017). *Knowledge Exchange Framework*. Retrieved September 22nd, 2018 from http://www.hefce.ac.uk/ke/kef

IBM. (2015). *The data center impact of cloud, analytics, mobile, social and security: a mandate for infrastructure agility, dynamic optimization and software defined environments.* IBM Global Technology Services. Thought leadership white paper.

Kao, R. (2006). Defining Entrepreneurship: Past and Present and? *Creativity and Innovation Management*, *2*(1), 69–70. doi:10.1111/j.1467-8691.1993.tb00073.x

Overli, H. (2018, September 9). For the good of us all, every business must be a technology business. *Sunday Telegraph Business and Money*, p. 2.

Stam, E., & Bosma, N. (2015). Growing Entrepreneurial Economies: Entrepreneurship and Regional Development. In T. Baker & F. Welter (Eds.), *The Routledge Companion to Entrepreneurship* (pp. 325-340). London: Routledge.

University of Gloucestershire. (2012). *Strategic Plan 2012-2017*. Retrieved September 19th, 2015 from http://www.glos.ac.uk/docs/download/Sustainability/Strategic-Plan-2012-2017.pdf

Wallace, T. (2017, November 13). UK firms missing out on £100bn by ignoring tried and tested tech, says CBI. *The Telegraph - Business.* Retrieved September 22nd, 2018 from https://www.telegraph.co.uk/business/2017/11/13/uk-firms-missing-100bn-ignoring-tried-tested-tech-says-cbi/

Wynn, M., & Jones, P. (2018). Context and Entrepreneurship in Knowledge Transfer Partnerships with Small Business Enterprises. *International Journal of Entrepreneurship and Innovation*, 1–13. Retrieved from journals.sagepub.com/home/iei

Zhang, Q. (2018). Theory, practice and policy: A longitudinal study of university knowledge exchange in the UK. *Industry and Higher Education*, *32*(2), 80–92. doi:10.1177/0950422218761500

Related Readings

To continue IGI Global's long-standing tradition of advancing innovation through emerging research, please find below a compiled list of recommended IGI Global book chapters and journal articles in the areas of higher education, business models, and organizational development. These related readings will provide additional information and guidance to further enrich your knowledge and assist you with your own research.

Adams, S., Klobodu, E. K., & Lamptey, R. O. (2017). Electric Power Transmission, Distribution Losses, and Economic Growth in Ghana. In R. Das (Ed.), *Social, Health, and Environmental Infrastructures for Economic Growth* (pp. 216–233). Hershey, PA: IGI Global. doi:10.4018/978-1-5225-2364-2.ch011

Adams, S., Klobodu, E. K., & Lamptey, R. O. (2017). Health Infrastructure and Economic Growth in Sub-Saharan Africa. In R. Das (Ed.), *Social, Health, and Environmental Infrastructures for Economic Growth* (pp. 82–98). Hershey, PA: IGI Global. doi:10.4018/978-1-5225-2364-2.ch005

Adetiloye, K. A., Eke, P. O., & Taiwo, J. N. (2017). Public-Sector Project Abandonment Decision: A Test of the Ricardian Equivalence Theory on the Failed Lagos Metroline in Nigeria. In R. Das (Ed.), *Social, Health, and Environmental Infrastructures for Economic Growth* (pp. 188–215). Hershey, PA: IGI Global. doi:10.4018/978-1-5225-2364-2.ch010

Adhikary, M., & Khatun, M. (2016). Issues of Convergence: Some Evidences of SAARC Countries. In R. Das (Ed.), *Handbook of Research on Global Indicators of Economic and Political Convergence* (pp. 119–143). Hershey, PA: IGI Global. doi:10.4018/978-1-5225-0215-9.ch006

Ahamer, G. (2016). Quality Assurance for a Developmental "Global Studies" (GS) Curriculum. In W. Nuninger & J. Châtelet (Eds.), *Handbook of Research on Quality Assurance and Value Management in Higher Education* (pp. 160–197). Hershey, PA: IGI Global. doi:10.4018/978-1-5225-0024-7.ch007

Ahamer, G. (2017). Learning and Space Mean Communication: Theories Rooting in China, India, and Europe. In S. Mukerji & P. Tripathi (Eds.), *Handbook of Research on Administration, Policy, and Leadership in Higher Education* (pp. 81–111). Hershey, PA: IGI Global. doi:10.4018/978-1-5225-0672-0.ch005

Akhter, T., & Palit, A. (2016). Effects of Neoliberal Globalization on Readymade Garment Industry of Bangladesh: A Case Study for Chittagong. In R. Das (Ed.), *Handbook of Research on Global Indicators of Economic and Political Convergence* (pp. 538–557). Hershey, PA: IGI Global. doi:10.4018/978-1-5225-0215-9.ch023

Akincilar, A. (2017). Do Judgmental Factors Lead to a Good Decision on Investing in a Currency or Mislead the Financial Player?: An Application in Turkey. In J. Stanković, P. Delias, S. Marinković, & S. Rochhia (Eds.), *Tools and Techniques for Economic Decision Analysis* (pp. 43–62). Hershey, PA: IGI Global. doi:10.4018/978-1-5225-0959-2.ch003

Alegre de la Rosa, O. M., & Angulo, L. M. (2017). Social Inclusion and Intercultural Values in a School of Education. In S. Mukerji & P. Tripathi (Eds.), *Handbook of Research on Administration, Policy, and Leadership in Higher Education* (pp. 518–531). Hershey, PA: IGI Global. doi:10.4018/978-1-5225-0672-0.ch020

Allegreni, F. (2017). Crowdfunding as a Marketing Tool. In W. Vassallo (Ed.), *Crowdfunding for Sustainable Entrepreneurship and Innovation* (pp. 187–203). Hershey, PA: IGI Global. doi:10.4018/978-1-5225-0568-6.ch011

Almaraz-Menéndez, F., & Maz-Machado, A. (2017). Organizing Digital Production in a Classic Higher Education Institution: The Case of the University of Salamanca. In S. Mukerji & P. Tripathi (Eds.), *Handbook of Research on Administration, Policy, and Leadership in Higher Education* (pp. 157–172). Hershey, PA: IGI Global. doi:10.4018/978-1-5225-0672-0.ch008

Altındağ, E. (2016). Current Approaches in Change Management. In A. Goksoy (Ed.), *Organizational Change Management Strategies in Modern Business* (pp. 24–51). Hershey, PA: IGI Global. doi:10.4018/978-1-4666-9533-7.ch002

Ambani, P. (2017). Crowdsourcing New Tools to Start Lean and Succeed in Entrepreneurship: Entrepreneurship in the Crowd Economy. In W. Vassallo (Ed.), *Crowdfunding for Sustainable Entrepreneurship and Innovation* (pp. 37–53). Hershey, PA: IGI Global. doi:10.4018/978-1-5225-0568-6.ch003

Amoako, G. K., Adjaison, G. K., & Osei-Bonsu, N. (2016). Role of Strategic Change Management in Emerging Markets: Ghanaian Perspective. In A. Goksoy (Ed.), *Organizational Change Management Strategies in Modern Business* (pp. 328–351). Hershey, PA: IGI Global. doi:10.4018/978-1-4666-9533-7.ch016

Andersson, B., & Mayer, T. (2017). University Rankings: Was Academic Life Perfect before Rankings? In K. Downing & F. Ganotice Jr., (Eds.), *World University Rankings and the Future of Higher Education* (pp. 70–86). Hershey, PA: IGI Global. doi:10.4018/978-1-5225-0819-9.ch004

Arsenijević, O. M., Orčić, D., & Kastratović, E. (2017). Development of an Optimization Tool for Intangibles in SMEs: A Case Study from Serbia with a Pilot Research in the Prestige by Milka Company. In M. Vemić (Ed.), *Optimal Management Strategies in Small and Medium Enterprises* (pp. 320–347). Hershey, PA: IGI Global. doi:10.4018/978-1-5225-1949-2.ch015

Baldock, C. (2017). Citations, Open Access and University Rankings. In K. Downing & F. Ganotice Jr., (Eds.), *World University Rankings and the Future of Higher Education* (pp. 129–139). Hershey, PA: IGI Global. doi:10.4018/978-1-5225-0819-9.ch007

Baporikar, N. (2016). Stakeholder Approach for Quality Higher Education. In W. Nuninger & J. Châtelet (Eds.), *Handbook of Research on Quality Assurance and Value Management in Higher Education* (pp. 1–26). Hershey, PA: IGI Global. doi:10.4018/978-1-5225-0024-7.ch001

Basu, I. (2017). Role of Infrastructure Development to Empower Women: An Over-Determined View. In R. Das (Ed.), *Social, Health, and Environmental Infrastructures for Economic Growth* (pp. 39–56). Hershey, PA: IGI Global. doi:10.4018/978-1-5225-2364-2.ch003

Ben Rejeb, W. (2017). Empirical Evidence on Corporate Governance Impact on CSR Disclosure in Developing Economies: The Tunisian and Egyptian Contexts. In D. Jamali (Ed.), *Comparative Perspectives on Global Corporate Social Responsibility* (pp. 116–137). Hershey, PA: IGI Global. doi:10.4018/978-1-5225-0720-8.ch006

Bisschoff, T. C., & Nieto, M. L. (2017). Leadership and Followership in Post-1992 University Business Schools in England. In S. Mukerji & P. Tripathi (Eds.), *Handbook of Research on Administration, Policy, and Leadership in Higher Education* (pp. 461–495). Hershey, PA: IGI Global. doi:10.4018/978-1-5225-0672-0.ch018

Biswas, R. (2017). "Theory of IRE with (α,β,γ) Norm": An Engineering Model for Higher Education Management (HEM) & Policy Administration in India. In S. Mukerji & P. Tripathi (Eds.), *Handbook of Research on Administration, Policy, and Leadership in Higher Education* (pp. 274–347). Hershey, PA: IGI Global. doi:10.4018/978-1-5225-0672-0.ch013

Blomme, R. J., & Morsch, J. (2016). Organizations as Social Networks: The Role of the Compliance Officer as Agent of Change in Implementing Rules and Codes of Conduct. In A. Goksoy (Ed.), *Organizational Change Management Strategies in Modern Business* (pp. 110–121). Hershey, PA: IGI Global. doi:10.4018/978-1-4666-9533-7.ch006

Boachie, C. (2017). Decision Making under Risk and Uncertainty in the Oil and Gas Industry. In J. Stanković, P. Delias, S. Marinković, & S. Rochhia (Eds.), *Tools and Techniques for Economic Decision Analysis* (pp. 181–207). Hershey, PA: IGI Global. doi:10.4018/978-1-5225-0959-2.ch009

Bridgeforth, J. S. (2017). Multicultural Leadership in Higher Education. In R. Styron Jr & J. Styron (Eds.), *Comprehensive Problem-Solving and Skill Development for Next-Generation Leaders* (pp. 139–164). Hershey, PA: IGI Global. doi:10.4018/978-1-5225-1968-3.ch007

Candelise, C. (2017). The Application of Crowdfunding to the Energy Sector. In W. Vassallo (Ed.), *Crowdfunding for Sustainable Entrepreneurship and Innovation* (pp. 266–287). Hershey, PA: IGI Global. doi:10.4018/978-1-5225-0568-6.ch015

Carvalho, G., Simões, M., & Duarte, A. P. (2017). Export-Led Recovery in Portugal: Can It Also Sustain Growth? In J. Stanković, P. Delias, S. Marinković, & S. Rochhia (Eds.), *Tools and Techniques for Economic Decision Analysis* (pp. 129–150). Hershey, PA: IGI Global. doi:10.4018/978-1-5225-0959-2.ch006

Chakrabarti, G., & Sen, C. (2016). Green Convergence in Emerging Nations: The Determinants and the Possibilities. In R. Das (Ed.), *Handbook of Research on Global Indicators of Economic and Political Convergence* (pp. 448–473). Hershey, PA: IGI Global. doi:10.4018/978-1-5225-0215-9.ch020

Chakraborty, S. (2016). Economic Convergence and Real Dimensions: The Case of Shelter Deprivation. In R. Das (Ed.), *Handbook of Research on Global Indicators of Economic and Political Convergence* (pp. 338–363). Hershey, PA: IGI Global. doi:10.4018/978-1-5225-0215-9.ch015

Chan, D., & Kumar, S. (2017). Social Media Strategies for Small and Medium Scale Enterprise in the Klang Valley Region of Malaysia. In N. Ahmad, T. Ramayah, H. Halim, & S. Rahman (Eds.), *Handbook of Research on Small and Medium Enterprises in Developing Countries* (pp. 377–400). Hershey, PA: IGI Global. doi:10.4018/978-1-5225-2165-5.ch018

Chatterjee, N. (2017). An Assessment of Infrastructural Facilities in the Dryland Areas of West Bengal. In R. Das (Ed.), *Social, Health, and Environmental Infrastructures for Economic Growth* (pp. 234–268). Hershey, PA: IGI Global. doi:10.4018/978-1-5225-2364-2.ch012

Chatterjee, N., & Dinda, S. (2016). Convergence of Forest Resources in Jangalmahal, West Bengal. In R. Das (Ed.), *Handbook of Research on Global Indicators of Economic and Political Convergence* (pp. 511–537). Hershey, PA: IGI Global. doi:10.4018/978-1-5225-0215-9.ch022

Chatterjee, T., & Dinda, S. (2016). Health Status and Convergence in Developing Open Economies: Is Health Status Converging in Developing Economies? In R. Das (Ed.), *Handbook of Research on Global Indicators of Economic and Political Convergence* (pp. 364–381). Hershey, PA: IGI Global. doi:10.4018/978-1-5225-0215-9.ch016

Chaudhury, A. R. (2016). How Does Inequality of Opportunity Persist from Education to Earning?: Some Recent Evidences from India. In R. Das (Ed.), *Handbook of Research on Global Indicators of Economic and Political Convergence* (pp. 314–337). Hershey, PA: IGI Global. doi:10.4018/978-1-5225-0215-9.ch014

Cinelli, S. A. (2017). Real Estate Crowdfunding: 2015 and Beyond. In W. Vassallo (Ed.), *Crowdfunding for Sustainable Entrepreneurship and Innovation* (pp. 238–265). Hershey, PA: IGI Global. doi:10.4018/978-1-5225-0568-6.ch014

Cinelli, S. A. (2017). The World's Oldest Profession - Now and Then: Disruption of the Commercial Banking Model. In W. Vassallo (Ed.), *Crowdfunding for Sustainable Entrepreneurship and Innovation* (pp. 78–89). Hershey, PA: IGI Global. doi:10.4018/978-1-5225-0568-6.ch005

Çolakoğlu, S., Chung, Y., & Tarhan, A. B. (2016). Strategic Human Resource Management in Facilitating Organizational Change. In A. Goksoy (Ed.), *Organizational Change Management Strategies in Modern Business* (pp. 172–192). Hershey, PA: IGI Global. doi:10.4018/978-1-4666-9533-7.ch009

Damnjanović, A. M. (2017). Knowledge Management Optimization through IT and E-Business Utilization: A Qualitative Study on Serbian SMEs. In M. Vemić (Ed.), *Optimal Management Strategies in Small and Medium Enterprises* (pp. 249–267). Hershey, PA: IGI Global. doi:10.4018/978-1-5225-1949-2.ch012

Daneshpour, H. (2017). Integrating Sustainable Development into Project Portfolio Management through Application of Open Innovation. In M. Vemić (Ed.), *Optimal Management Strategies in Small and Medium Enterprises* (pp. 370–387). Hershey, PA: IGI Global. doi:10.4018/978-1-5225-1949-2.ch017

Das, R. C., Das, A., & Martin, F. (2016). Convergence Analysis of Households' Consumption Expenditure: A Cross Country Study. In R. Das (Ed.), *Handbook of Research on Global Indicators of Economic and Political Convergence* (pp. 1–28). Hershey, PA: IGI Global. doi:10.4018/978-1-5225-0215-9.ch001

Das, U., Das, R. C., & Ray, K. (2016). Convergence and Equality of Road Infrastructure: A Cross Country Analysis. In R. Das (Ed.), *Handbook of Research on Global Indicators of Economic and Political Convergence* (pp. 170–183). Hershey, PA: IGI Global. doi:10.4018/978-1-5225-0215-9.ch008

Dastaviz, A. H. (2016). Integration of Competing Values and Knowledge Organisational Activities in a New Model. *International Journal of Knowledge-Based Organizations*, *6*(4), 50–61. doi:10.4018/IJKBO.2016100105

Datta, P. K., Tiwary, H., & Chakrabarti, S. (2017). MGNREGS of India: Complementarities Between Employment and Infrastructure. In R. Das (Ed.), *Social, Health, and Environmental Infrastructures for Economic Growth* (pp. 57–81). Hershey, PA: IGI Global. doi:10.4018/978-1-5225-2364-2.ch004

Davidson, D. K., & Yin, J. (2017). Corporate Social Responsibility (CSR) in China: A Contextual Exploration. In D. Jamali (Ed.), *Comparative Perspectives on Global Corporate Social Responsibility* (pp. 28–48). Hershey, PA: IGI Global. doi:10.4018/978-1-5225-0720-8.ch002

de Burgh-Woodman, H., Bressan, A., & Torrisi, A. (2017). An Evaluation of the State of the CSR Field in Australia: Perspectives from the Banking and Mining Sectors. In D. Jamali (Ed.), *Comparative Perspectives on Global Corporate Social Responsibility* (pp. 138–164). Hershey, PA: IGI Global. doi:10.4018/978-1-5225-0720-8.ch007

De Ruiter, M., Blomme, R. J., & Schalk, R. (2016). Reducing the Negative Effects of Psychological Contract Breach during Management-Imposed Change: A Trickle-Down Model of Management Practices. In A. Goksoy (Ed.), *Organizational Change Management Strategies in Modern Business* (pp. 122–142). Hershey, PA: IGI Global. doi:10.4018/978-1-4666-9533-7.ch007

Dedousis, E., & Rutter, R. N. (2016). Workforce Localisation and Change Management: The View from the Gulf. In A. Goksoy (Ed.), *Organizational Change Management Strategies in Modern Business* (pp. 301–327). Hershey, PA: IGI Global. doi:10.4018/978-1-4666-9533-7.ch015

Demiray, M., & Aslanbay, Y. (2017). The Crowdfunding Communities and the Value of Identification for Sustainability of Co-Creation. In W. Vassallo (Ed.), *Crowdfunding for Sustainable Entrepreneurship and Innovation* (pp. 155–174). Hershey, PA: IGI Global. doi:10.4018/978-1-5225-0568-6.ch009

Demiray, M., Burnaz, S., & Aslanbay, Y. (2017). The Crowdfunding Market, Models, Platforms, and Projects. In W. Vassallo (Ed.), *Crowdfunding for Sustainable Entrepreneurship and Innovation* (pp. 90–126). Hershey, PA: IGI Global. doi:10.4018/978-1-5225-0568-6.ch006

Dikeç, A., Kane, V., & Çapar, N. (2017). Cross-Country and Cross-Sector CSR Variations: A Comparative Analysis of CSR Reporting in the U.S., South Korea, and Turkey. In D. Jamali (Ed.), *Comparative Perspectives on Global Corporate Social Responsibility* (pp. 69–95). Hershey, PA: IGI Global. doi:10.4018/978-1-5225-0720-8.ch004

Dixon, K. V., Gonzalez-Carriedo, R., & Dixon-Krauss, L. (2017). The Implementation and Analysis of an International Student Teaching Program. In L. Leavitt, S. Wisdom, & K. Leavitt (Eds.), *Cultural Awareness and Competency Development in Higher Education* (pp. 168–188). Hershey, PA: IGI Global. doi:10.4018/978-1-5225-2145-7.ch010

Duruji, M. M., & Urenma, D. F. (2016). The Environmentalism and Politics of Climate Change: A Study of the Process of Global Convergence through UNFCCC Conferences. In R. Das (Ed.), *Handbook of Research on Global Indicators of Economic and Political Convergence* (pp. 398-429). Hershey, PA: IGI Global. doi:10.4018/978-1-5225-0215-9.ch018

Eldin, E. M. (2016). The Impact of Cost Sharing on Quality of Education in Egypt's Public Universities. In W. Nuninger & J. Châtelet (Eds.), *Handbook of Research on Quality Assurance and Value Management in Higher Education* (pp. 58–84). Hershey, PA: IGI Global. doi:10.4018/978-1-5225-0024-7.ch003

Elufiede, O. J., & Flynn, B. B. (2017). Mentor the Leader: A Transformational Approach. In R. Styron Jr & J. Styron (Eds.), *Comprehensive Problem-Solving and Skill Development for Next-Generation Leaders* (pp. 188–209). Hershey, PA: IGI Global. doi:10.4018/978-1-5225-1968-3.ch009

Erne, R. (2016). Change Management Revised. In A. Goksoy (Ed.), *Organizational Change Management Strategies in Modern Business* (pp. 1–23). Hershey, PA: IGI Global. doi:10.4018/978-1-4666-9533-7.ch001

Ertürk, E., Yılmaz, D., & Çetin, I. (2016). Optimum Currency Area Theory and Business Cycle Convergence in EMU: Considering the Sovereign Debt Crisis. In R. Das (Ed.), *Handbook of Research on Global Indicators of Economic and Political Convergence* (pp. 67–91). Hershey, PA: IGI Global. doi:10.4018/978-1-5225-0215-9.ch004

Eryılmaz, M. E., Erdur, D. A., Bektaş, O., Kara, E., & Aydoğan, E. (2016). The Case of ISO 9000 Quality Management System Certification in a Faculty of a Turkish Public University: Triggers, Processes, and Consequences. In W. Nuninger & J. Châtelet (Eds.), *Handbook of Research on Quality Assurance and Value Management in Higher Education* (pp. 257–278). Hershey, PA: IGI Global. doi:10.4018/978-1-5225-0024-7.ch010

Eryılmaz, M. E., & Eryılmaz, F. (2016). Change Emphasis in Mission and Vision Statements of the First 1000 Turkish Organizations: A Content Analysis. In A. Goksoy (Ed.), *Organizational Change Management Strategies in Modern Business* (pp. 352–362). Hershey, PA: IGI Global. doi:10.4018/978-1-4666-9533-7.ch017

Frączkiewicz-Wronka, A., Tkacz, M., & Arando, S. (2017). The Business Model of a Public Social Partnership: Contextual Determinants. In M. Lewandowski & B. Kożuch (Eds.), *Public Sector Entrepreneurship and the Integration of Innovative Business Models* (pp. 108–138). Hershey, PA: IGI Global. doi:10.4018/978-1-5225-2215-7.ch005

Friedrich, P., & Chebotareva, M. (2017). Options for Applying Functional Overlapping Competing Jurisdictions (FOCJs) for Municipal Cooperation in Russia. In M. Lewandowski & B. Kożuch (Eds.), *Public Sector Entrepreneurship and the Integration of Innovative Business Models* (pp. 73–107). Hershey, PA: IGI Global. doi:10.4018/978-1-5225-2215-7.ch004

Gan, P. L., Mahmud, I., Ramayah, T., & Zuhora, F. T. (2017). Understanding Attitude towards Green IT among Professionals in IT Service SMEs in Bangladesh. In N. Ahmad, T. Ramayah, H. Halim, & S. Rahman (Eds.), *Handbook of Research on Small and Medium Enterprises in Developing Countries* (pp. 48–66). Hershey, PA: IGI Global. doi:10.4018/978-1-5225-2165-5.ch003

Ganesan, Y., Haron, H., Amran, A., & Ooi, S. K. (2017). Corporate Social Responsibility in SMEs: The Role of Non-Audit Services. In N. Ahmad, T. Ramayah, H. Halim, & S. Rahman (Eds.), *Handbook of Research on Small and Medium Enterprises in Developing Countries* (pp. 345–359). Hershey, PA: IGI Global. doi:10.4018/978-1-5225-2165-5.ch016

Ghosh, D., & Dinda, S. (2017). Health Infrastructure and Economic Development in India. In R. Das (Ed.), *Social, Health, and Environmental Infrastructures for Economic Growth* (pp. 99–119). Hershey, PA: IGI Global. doi:10.4018/978-1-5225-2364-2.ch006

Ghosh, S. (2017). Changes in Infrastructural Condition of Slums in India: A State Level Analysis. In R. Das (Ed.), *Social, Health, and Environmental Infrastructures for Economic Growth* (pp. 14–38). Hershey, PA: IGI Global. doi:10.4018/978-1-5225-2364-2.ch002

Gonzalez, R. V. (2016). Knowledge Management Process in Multi-Site Provision of Service. *International Journal of Knowledge Management*, *12*(2), 20–37. doi:10.4018/IJKM.2016040102

Greco, T. (2017). Horizontal Crowdfunding Platforms: Socio-Economic Impact and Exogenous/Endogenous System of Rules. In W. Vassallo (Ed.), *Crowdfunding for Sustainable Entrepreneurship and Innovation* (pp. 127–139). Hershey, PA: IGI Global. doi:10.4018/978-1-5225-0568-6.ch007

Greco, T. (2017). Vertical and Community-Based Crowdfunding Platforms. In W. Vassallo (Ed.), *Crowdfunding for Sustainable Entrepreneurship and Innovation* (pp. 140–154). Hershey, PA: IGI Global. doi:10.4018/978-1-5225-0568-6.ch008

Hack-Polay, D., & Qiu, H. (2017). Doing Good Doing Well: Discussion of CSR in the Pulp and Paper Industry in the Asian Context. In D. Jamali (Ed.), *Comparative Perspectives on Global Corporate Social Responsibility* (pp. 226–240). Hershey, PA: IGI Global. doi:10.4018/978-1-5225-0720-8.ch011

Halim, H. A., Ahmad, N. H., Hanifah, H., & Ramayah, T. (2017). The Impediments to Entrepreneurial Ventures among the Bottom of Pyramid Community in Northern Malaysia. In N. Ahmad, T. Ramayah, H. Halim, & S. Rahman (Eds.), *Handbook of Research on Small and Medium Enterprises in Developing Countries* (pp. 67–81). Hershey, PA: IGI Global. doi:10.4018/978-1-5225-2165-5.ch004

Hanifah, H., Halim, H. A., Ahmad, N. H., & Vafaei-Zadeh, A. (2017). Innovation Culture as a Mediator Between Specific Human Capital and Innovation Performance Among Bumiputera SMEs in Malaysia. In N. Ahmad, T. Ramayah, H. Halim, & S. Rahman (Eds.), *Handbook of Research on Small and Medium Enterprises in Developing Countries* (pp. 261–279). Hershey, PA: IGI Global. doi:10.4018/978-1-5225-2165-5.ch012

Haro-de-Rosario, A., del Mar Gálvez-Rodríguez, M., & Caba-Pérez, M. D. (2017). Determinants of Corporate Social Responsibility Disclosure in Latin American Companies: An Analysis of the Oil and Gas Sector. In D. Jamali (Ed.), *Comparative Perspectives on Global Corporate Social Responsibility* (pp. 165–184). Hershey, PA: IGI Global. doi:10.4018/978-1-5225-0720-8.ch008

Hassaan, M. (2016). Sustainability Reporting in Transitional Economies. In R. Das (Ed.), *Handbook of Research on Global Indicators of Economic and Political Convergence* (pp. 184–204). Hershey, PA: IGI Global. doi:10.4018/978-1-5225-0215-9.ch009

Hassan, A., & Lund-Thomsen, P. (2017). Multi-Stakeholder Initiatives and Corporate Social Responsibility in Global Value Chains: Towards an Analytical Framework and a Methodology. In D. Jamali (Ed.), *Comparative Perspectives on Global Corporate Social Responsibility* (pp. 241–257). Hershey, PA: IGI Global. doi:10.4018/978-1-5225-0720-8.ch012

Hossain, M. M., & Ibrahim, Y. (2017). Working Capital Financing by Banks in Small Enterprises: Problems and Challenges for Bangladesh. In N. Ahmad, T. Ramayah, H. Halim, & S. Rahman (Eds.), *Handbook of Research on Small and Medium Enterprises in Developing Countries* (pp. 137–158). Hershey, PA: IGI Global. doi:10.4018/978-1-5225-2165-5.ch007

Indiran, L., Khalifah, Z., Ismail, K., & Ramanathan, S. (2017). Business Incubation in Malaysia: An Overview of Multimedia Super Corridor, Small and Medium Enterprises, and Incubators in Malaysia. In N. Ahmad, T. Ramayah, H. Halim, & S. Rahman (Eds.), *Handbook of Research on Small and Medium Enterprises in Developing Countries* (pp. 322–344). Hershey, PA: IGI Global. doi:10.4018/978-1-5225-2165-5.ch015

Irizar, J., & Wynn, M. (2018). A new model for improving project risk management in the automotive industry. *International Journal of Risk and Contingency Management*, *7*(3), 53–72. doi:10.4018/IJRCM.2018070103

Issa, T., & Pick, D. (2017). Teaching Business Ethics Post GFC: A Corporate Social Responsibility of Universities. In D. Jamali (Ed.), *Comparative Perspectives on Global Corporate Social Responsibility* (pp. 290–307). Hershey, PA: IGI Global. doi:10.4018/978-1-5225-0720-8.ch015

Ivana, B., & Vrkić, F. (2017). Crisis Communication and Crisis Management during the Crisis: Case Study of Croatia. In J. Stanković, P. Delias, S. Marinković, & S. Rochhia (Eds.), *Tools and Techniques for Economic Decision Analysis* (pp. 208–224). Hershey, PA: IGI Global. doi:10.4018/978-1-5225-0959-2.ch010

Ivančević, K. (2017). Insurance as an Optimization Tool for Risk Management in Small and Medium-Sized Enterprises. In M. Vemić (Ed.), *Optimal Management Strategies in Small and Medium Enterprises* (pp. 178–197). Hershey, PA: IGI Global. doi:10.4018/978-1-5225-1949-2.ch009

Jain, P. (2017). A Crowd-Funder Value (CFV) Framework for Crowd-Investment: A Roadmap for Entrepreneurial Success in the Contemporary Society. In W. Vassallo (Ed.), *Crowdfunding for Sustainable Entrepreneurship and Innovation* (pp. 288–309). Hershey, PA: IGI Global. doi:10.4018/978-1-5225-0568-6.ch016

Jalali, A., Thurasamy, R., & Jaafar, M. (2017). The Moderating Effect of Social Capital in Relation to Entrepreneurial Orientation and Firm Performance. In N. Ahmad, T. Ramayah, H. Halim, & S. Rahman (Eds.), *Handbook of Research on Small and Medium Enterprises in Developing Countries* (pp. 82–115). Hershey, PA: IGI Global. doi:10.4018/978-1-5225-2165-5.ch005

Jana, S. K., & Karmakar, A. K. (2016). Food Security in Asia: Is There Convergence? In R. Das (Ed.), *Handbook of Research on Global Indicators of Economic and Political Convergence* (pp. 382–397). Hershey, PA: IGI Global. doi:10.4018/978-1-5225-0215-9.ch017

Jana, S. K., & Karmakar, A. K. (2017). Infrastructure, Education, and Economic Development in India: A State Level Analysis. In R. Das (Ed.), *Social, Health, and Environmental Infrastructures for Economic Growth* (pp. 1–13). Hershey, PA: IGI Global. doi:10.4018/978-1-5225-2364-2.ch001

Jankovic-Milic, V., & Džunić, M. (2017). Measuring Governance: The Application of Grey Relational Analysis on World Governance Indicators. In J. Stanković, P. Delias, S. Marinković, & S. Rochhia (Eds.), *Tools and Techniques for Economic Decision Analysis* (pp. 104–128). Hershey, PA: IGI Global. doi:10.4018/978-1-5225-0959-2.ch005

Jayapal, P., & Omar, A. (2017). The Role of Value Co-Creation on Brand Image: A Conceptual Framework for the Market Performance of SMEs in Malaysia. In N. Ahmad, T. Ramayah, H. Halim, & S. Rahman (Eds.), *Handbook of Research on Small and Medium Enterprises in Developing Countries* (pp. 185–207). Hershey, PA: IGI Global. doi:10.4018/978-1-5225-2165-5.ch009

Kaplan, J., & Montiel, I. (2017). East vs. West Approaches to Reporting Corporate Sustainability Strategies to the World: Corporate Sustainability Reporting: East vs. West. In D. Jamali (Ed.), *Comparative Perspectives on Global Corporate Social Responsibility* (pp. 49–68). Hershey, PA: IGI Global. doi:10.4018/978-1-5225-0720-8.ch003

Kasemsap, K. (2016). The Roles of Organizational Change Management and Resistance to Change in the Modern Business World. In A. Goksoy (Ed.), *Organizational Change Management Strategies in Modern Business* (pp. 143–171). Hershey, PA: IGI Global. doi:10.4018/978-1-4666-9533-7.ch008

Kasemsap, K. (2017). Encouraging Internationalization and Entrepreneurial Orientation in Small and Medium Enterprises. In N. Ahmad, T. Ramayah, H. Halim, & S. Rahman (Eds.), *Handbook of Research on Small and Medium Enterprises in Developing Countries* (pp. 233–259). Hershey, PA: IGI Global. doi:10.4018/978-1-5225-2165-5.ch011

Kesti, M. O., Leinonen, J., & Kesti, T. (2017). The Productive Leadership Game: From Theory to Game-Based Learning. In M. Lewandowski & B. Kożuch (Eds.), *Public Sector Entrepreneurship and the Integration of Innovative Business Models* (pp. 238–260). Hershey, PA: IGI Global. doi:10.4018/978-1-5225-2215-7.ch010

King, D. R. (2016). Management as a Limit to Organizational Change: Implications for Acquisitions. In A. Goksoy (Ed.), *Organizational Change Management Strategies in Modern Business* (pp. 52–73). Hershey, PA: IGI Global. doi:10.4018/978-1-4666-9533-7.ch003

Kishna, T., Blomme, R. J., & van der Veen, J. A. (2016). Organizational Routines: Developing a Duality Model to Explain the Effects of Strategic Change Initiatives. In A. Goksoy (Ed.), *Organizational Change Management Strategies in Modern Business* (pp. 363–385). Hershey, PA: IGI Global. doi:10.4018/978-1-4666-9533-7.ch018

Kleverlaan, R. (2017). How to Launch a Successful Crowdfunding Campaign. In W. Vassallo (Ed.), *Crowdfunding for Sustainable Entrepreneurship and Innovation* (pp. 224–237). Hershey, PA: IGI Global. doi:10.4018/978-1-5225-0568-6.ch013

Kożuch, B., & Jabłoński, A. (2017). Adopting the Concept of Business Models in Public Management. In M. Lewandowski & B. Kożuch (Eds.), *Public Sector Entrepreneurship and the Integration of Innovative Business Models* (pp. 10–46). Hershey, PA: IGI Global. doi:10.4018/978-1-5225-2215-7.ch002

Laha, A. (2016). Association between Governance and Human Development in South Asia: A Cross Country Analysis. In R. Das (Ed.), *Handbook of Research on Global Indicators of Economic and Political Convergence* (pp. 254–273). Hershey, PA: IGI Global. doi:10.4018/978-1-5225-0215-9.ch012

Lavassani, K. M., & Movahedi, B. (2017). Applications Driven Information Systems: Beyond Networks toward Business Ecosystems. *International Journal of Innovation in the Digital Economy*, *8*(1), 61–75. doi:10.4018/IJIDE.2017010104

Lerro, A. M. (2017). Legal Aspects and Regulation in Crowdfunding: Comparisons across Countries. In W. Vassallo (Ed.), *Crowdfunding for Sustainable Entrepreneurship and Innovation* (pp. 204–223). Hershey, PA: IGI Global. doi:10.4018/978-1-5225-0568-6.ch012

Lewandowski, M. (2017). Public Organizations and Business Model Innovation: The Role of Public Service Design. In M. Lewandowski & B. Kożuch (Eds.), *Public Sector Entrepreneurship and the Integration of Innovative Business Models* (pp. 47–72). Hershey, PA: IGI Global. doi:10.4018/978-1-5225-2215-7.ch003

Lousã, M. D., & Monteiro, J. A. (2015). Explanatory Model of Adoption, Development and Utilization of Administrative Workflow Systems. *International Journal of Sociotechnology and Knowledge Development*, *7*(4), 31–52. doi:10.4018/IJSKD.2015100103

Maciejczak, M., & Słodki, A. (2017). Initial Price Strategies of Polish Micro and Small Enterprises: An Application of Game Theory for Industrial Organization of the SME Sector. In M. Vemić (Ed.), *Optimal Management Strategies in Small and Medium Enterprises* (pp. 126–143). Hershey, PA: IGI Global. doi:10.4018/978-1-5225-1949-2.ch007

Magala, S. J. (2017). Between Davos and Porto Alegre: Democratic Entrepreneurship as Crowdsourcing for Ideas. In M. Lewandowski & B. Kożuch (Eds.), *Public Sector Entrepreneurship and the Integration of Innovative Business Models* (pp. 1–9). Hershey, PA: IGI Global. doi:10.4018/978-1-5225-2215-7.ch001

Malega, P. (2017). Small and Medium Enterprises in the Slovak Republic: Status and Competitiveness of SMEs in the Global Markets and Possibilities of Optimization. In M. Vemić (Ed.), *Optimal Management Strategies in Small and Medium Enterprises* (pp. 102–124). Hershey, PA: IGI Global. doi:10.4018/978-1-5225-1949-2.ch006

Malek, N. A., & Takala, J. A. (2017). Regression Analysis for Environmental Practices: Participation among Finnish SMEs. In M. Vemić (Ed.), *Optimal Management Strategies in Small and Medium Enterprises* (pp. 44–59). Hershey, PA: IGI Global. doi:10.4018/978-1-5225-1949-2.ch003

Malik, A. (2016). The Role of HR Strategies in Change. In A. Goksoy (Ed.), *Organizational Change Management Strategies in Modern Business* (pp. 193–215). Hershey, PA: IGI Global. doi:10.4018/978-1-4666-9533-7.ch010

Mandal, C., & Gupta, A. C. (2016). A Comparative Study on World-Wide Carbon Emission Convergence: An Empirical Analysis. In R. Das (Ed.), *Handbook of Research on Global Indicators of Economic and Political Convergence* (pp. 430–447). Hershey, PA: IGI Global. doi:10.4018/978-1-5225-0215-9.ch019

Manzoor, A. (2017). Sustainable Infrastructures: A New Infrastructure Investment Strategy. In R. Das (Ed.), *Social, Health, and Environmental Infrastructures for Economic Growth* (pp. 145–164). Hershey, PA: IGI Global. doi:10.4018/978-1-5225-2364-2.ch008

Mazumdar, D. (2016). The Problems of Development Gap between Developed and Developing Nations: Is There Any Sign of Convergence? In R. Das (Ed.), *Handbook of Research on Global Indicators of Economic and Political Convergence* (pp. 29–50). Hershey, PA: IGI Global. doi:10.4018/978-1-5225-0215-9.ch002

Milošević, O. (2017). Critical Review of SME Regulation Optimization in Serbia: A Reflection on Harmonization with the EU Acquis. In M. Vemić (Ed.), *Optimal Management Strategies in Small and Medium Enterprises* (pp. 79–101). Hershey, PA: IGI Global. doi:10.4018/978-1-5225-1949-2.ch005

Molnar, R. M. (2017). Structural Change Management for Sustainable SME Development: Applying Classical Management Tools. In M. Vemić (Ed.), *Optimal Management Strategies in Small and Medium Enterprises* (pp. 348–369). Hershey, PA: IGI Global. doi:10.4018/978-1-5225-1949-2.ch016

Monahov, A. (2017). The Effects of Prudential Supervision on Bank Resiliency and Profits in a Multi-Agent Setting. In J. Stanković, P. Delias, S. Marinković, & S. Rochhia (Eds.), *Tools and Techniques for Economic Decision Analysis* (pp. 63–103). Hershey, PA: IGI Global. doi:10.4018/978-1-5225-0959-2.ch004

Mukherjee, S. (2017). Anatomy and Significance of Public Healthcare Expenditure and Economic Growth Nexus in India: Its Implications for Public Health Infrastructure Thereof. In R. Das (Ed.), *Social, Health, and Environmental Infrastructures for Economic Growth* (pp. 120–144). Hershey, PA: IGI Global. doi:10.4018/978-1-5225-2364-2.ch007

Mukhopadhyay, D., & Mandal, A. K. (2017). Infrastructure Development and Changing Market Penetration of Consumer Durables in Rural India: An Empirical Investigation. In R. Das (Ed.), *Social, Health, and Environmental Infrastructures for Economic Growth* (pp. 269–289). Hershey, PA: IGI Global. doi:10.4018/978-1-5225-2364-2.ch013

Mukhopadhyay, P., Sinha, M., & Sengupta, P. P. (2017). Importance of Sustainable Rural Development through Agrarian Reforms: An Indian Scenario. In R. Das (Ed.), *Social, Health, and Environmental Infrastructures for Economic Growth* (pp. 290–306). Hershey, PA: IGI Global. doi:10.4018/978-1-5225-2364-2.ch014

Musa, E. (2017). Crowdsourcing Social Innovation: Towards a Collaborative Social Capitalism. In W. Vassallo (Ed.), *Crowdfunding for Sustainable Entrepreneurship and Innovation* (pp. 54–77). Hershey, PA: IGI Global. doi:10.4018/978-1-5225-0568-6.ch004

Nekaj, E. L. (2017). The Crowd Economy: From the Crowd to Businesses to Public Administrations and Multinational Companies. In W. Vassallo (Ed.), *Crowdfunding for Sustainable Entrepreneurship and Innovation* (pp. 1–19). Hershey, PA: IGI Global. doi:10.4018/978-1-5225-0568-6.ch001

Neogi, C. (2016). Productivity Convergence and Asian Trade Blocks. In R. Das (Ed.), *Handbook of Research on Global Indicators of Economic and Political Convergence* (pp. 92–118). Hershey, PA: IGI Global. doi:10.4018/978-1-5225-0215-9.ch005

Niederer, P. M. (2017). Principles in Crowdfunding Benefits and Validation Tools: How to Ensure Campaign Success. In W. Vassallo (Ed.), *Crowdfunding for Sustainable Entrepreneurship and Innovation* (pp. 175–186). Hershey, PA: IGI Global. doi:10.4018/978-1-5225-0568-6.ch010

Omiunu, O. G. (2017). Deploying ICT to Enhance Small Businesses and Achieving Sustainable Development: A Paradigm to Reducing Poverty and Unemployment and Enhancing a Sustainable National Development in Nigeria. In N. Ahmad, T. Ramayah, H. Halim, & S. Rahman (Eds.), *Handbook of Research on Small and Medium Enterprises in Developing Countries* (pp. 208–232). Hershey, PA: IGI Global. doi:10.4018/978-1-5225-2165-5.ch010

Osman, M. N. (2017). Internet-Based Social Reporting in Emerging Economies: Insights from Public Banks in Egypt and the UAE. In D. Jamali (Ed.), *Comparative Perspectives on Global Corporate Social Responsibility* (pp. 96–115). Hershey, PA: IGI Global. doi:10.4018/978-1-5225-0720-8.ch005

Özgeldi, M. (2016). Role of Human Resources in Change. In A. Goksoy (Ed.), *Organizational Change Management Strategies in Modern Business* (pp. 216–229). Hershey, PA: IGI Global. doi:10.4018/978-1-4666-9533-7.ch011

Patra, S. (2016). Role of SAARC in Convergence of South Asian Economies. In R. Das (Ed.), *Handbook of Research on Global Indicators of Economic and Political Convergence* (pp. 144–169). Hershey, PA: IGI Global. doi:10.4018/978-1-5225-0215-9.ch007

Patra, S. (2017). Energy Challenges and Infrastructure Development in South Asia. In R. Das (Ed.), *Social, Health, and Environmental Infrastructures for Economic Growth* (pp. 165–187). Hershey, PA: IGI Global. doi:10.4018/978-1-5225-2364-2.ch009

Puaschunder, J. (2017). The Call for Global Responsible Inter-Generational Leadership: The Quest of an Integration of Inter-Generational Equity in Corporate Social Responsibility (CSR) Models. In D. Jamali (Ed.), *Comparative Perspectives on Global Corporate Social Responsibility* (pp. 276–289). Hershey, PA: IGI Global. doi:10.4018/978-1-5225-0720-8.ch014

Radosavljevic, M., & Andjelkovic, A. (2017). Multi-Criteria Decision Making Approach for Choosing Business Process for the Improvement: Upgrading of the Six Sigma Methodology. In J. Stanković, P. Delias, S. Marinković, & S. Rochhia (Eds.), *Tools and Techniques for Economic Decision Analysis* (pp. 225–247). Hershey, PA: IGI Global. doi:10.4018/978-1-5225-0959-2.ch011

Rahman, S. A., Ahmad, N. H., & Taghizadeh, S. K. (2017). On the Road to SME Sector Development in Bangladesh: A Guideline Based on Current Challenges and Opportunities. In N. Ahmad, T. Ramayah, H. Halim, & S. Rahman (Eds.), *Handbook of Research on Small and Medium Enterprises in Developing Countries* (pp. 117–136). Hershey, PA: IGI Global. doi:10.4018/978-1-5225-2165-5.ch006

Raimi, L. (2017). Leveraging CSR as a 'support-aid' for Triple Bottom-Line Development in Nigeria: Evidence from the Telecommunication Industry. In D. Jamali (Ed.), *Comparative Perspectives on Global Corporate Social Responsibility* (pp. 208–225). Hershey, PA: IGI Global. doi:10.4018/978-1-5225-0720-8.ch010

Rajamanickam, S. (2016). Exploring Landscapes in Regional Convergence: Environment and Sustainable Development in South Asia. In R. Das (Ed.), *Handbook of Research on Global Indicators of Economic and Political Convergence* (pp. 474–510). Hershey, PA: IGI Global. doi:10.4018/978-1-5225-0215-9.ch021

Ray, K., Das, R. C., & Das, U. (2016). Convergence Aspect of Capital Formation: A Study on Major Countries. In R. Das (Ed.), *Handbook of Research on Global Indicators of Economic and Political Convergence* (pp. 51–66). Hershey, PA: IGI Global. doi:10.4018/978-1-5225-0215-9.ch003

Richet, J. (2016). Internal Communication Failure in Times of Change. In A. Goksoy (Ed.), *Organizational Change Management Strategies in Modern Business* (pp. 289–300). Hershey, PA: IGI Global. doi:10.4018/978-1-4666-9533-7.ch014

Riemann, U. (2016). The Power of Three: A Blended Approach of Project-, Change Management, and Design Thinking. In A. Goksoy (Ed.), *Organizational Change Management Strategies in Modern Business* (pp. 74–94). Hershey, PA: IGI Global. doi:10.4018/978-1-4666-9533-7.ch004

Rossetti di Valdalbero, D., & Birnbaum, B. (2017). Towards a New Economy: Co-Creation and Open Innovation in a Trustworthy Europe. In W. Vassallo (Ed.), *Crowdfunding for Sustainable Entrepreneurship and Innovation* (pp. 20–36). Hershey, PA: IGI Global. doi:10.4018/978-1-5225-0568-6.ch002

Saglietto, L., David, D., & Cezanne, C. (2017). Rethinking Social Capital Measurement. In J. Stanković, P. Delias, S. Marinković, & S. Rochhia (Eds.), *Tools and Techniques for Economic Decision Analysis* (pp. 248–268). Hershey, PA: IGI Global. doi:10.4018/978-1-5225-0959-2.ch012

Sanchez-Ruiz, L., & Blanco, B. (2017). Process Management for SMEs: Barriers, Enablers, and Benefits. In M. Vemić (Ed.), *Optimal Management Strategies in Small and Medium Enterprises* (pp. 293–319). Hershey, PA: IGI Global. doi:10.4018/978-1-5225-1949-2.ch014

Savic, M., & Zubovic, J. (2017). Modeling Labor Market Flows on the Basis of Sectoral Employment in Europe. In J. Stanković, P. Delias, S. Marinković, & S. Rochhia (Eds.), *Tools and Techniques for Economic Decision Analysis* (pp. 151–169). Hershey, PA: IGI Global. doi:10.4018/978-1-5225-0959-2.ch007

Scholtz, B., Koorsse, M., & Loleka, S. (2017). Understanding and Adoption of E-Finance in Small and Medium Enterprises (SMEs) in Developing Countries: A Study of Bangladesh and South Africa. In N. Ahmad, T. Ramayah, H. Halim, & S. Rahman (Eds.), *Handbook of Research on Small and Medium Enterprises in Developing Countries* (pp. 159–184). Hershey, PA: IGI Global. doi:10.4018/978-1-5225-2165-5.ch008

Šebestová, J., & Palová, Z. (2017). Support of Social Innovations: Case of the Czech Republic. In M. Lewandowski & B. Kożuch (Eds.), *Public Sector Entrepreneurship and the Integration of Innovative Business Models* (pp. 165–187). Hershey, PA: IGI Global. doi:10.4018/978-1-5225-2215-7.ch007

Seliga, R. (2017). Public Sector Marketing in Poland. In M. Lewandowski & B. Kożuch (Eds.), *Public Sector Entrepreneurship and the Integration of Innovative Business Models* (pp. 214–237). Hershey, PA: IGI Global. doi:10.4018/978-1-5225-2215-7.ch009

Šević, A., & Marinković, S. (2017). Investment Decision Making: Where Do We Stand? In J. Stanković, P. Delias, S. Marinković, & S. Rochhia (Eds.), *Tools and Techniques for Economic Decision Analysis* (pp. 1–23). Hershey, PA: IGI Global. doi:10.4018/978-1-5225-0959-2.ch001

Şimşit, Z. T., Günay, N. S., & Vayvay, Ö. (2016). Organizational Learning to Managing Change: Key Player of Continuous Improvement in the 21st Century. In A. Goksoy (Ed.), *Organizational Change Management Strategies in Modern Business* (pp. 95–109). Hershey, PA: IGI Global. doi:10.4018/978-1-4666-9533-7.ch005

Širůček, M., & Křen, L. (2017). Application of Markowitz Portfolio Theory by Building Optimal Portfolio on the US Stock Market. In J. Stanković, P. Delias, S. Marinković, & S. Rochhia (Eds.), *Tools and Techniques for Economic Decision Analysis* (pp. 24–42). Hershey, PA: IGI Global. doi:10.4018/978-1-5225-0959-2.ch002

Soniewicki, M. (2017). The Importance of Market Orientation in Creating a Competitive Advantage of Micro, Small, and Medium-Sized Companies in the Internationalization Process. In M. Vemić (Ed.), *Optimal Management Strategies in Small and Medium Enterprises* (pp. 1–21). Hershey, PA: IGI Global. doi:10.4018/978-1-5225-1949-2.ch001

Staszewska, B. M. (2017). Local Public Enterprise Business Model as Multiple Value Creation System. In M. Lewandowski & B. Kożuch (Eds.), *Public Sector Entrepreneurship and the Integration of Innovative Business Models* (pp. 188–213). Hershey, PA: IGI Global. doi:10.4018/978-1-5225-2215-7.ch008

Stefanova, J. S., & Wenner, Z. (2017). Entrepreneurship and Enterprise Value Creation in Support of Smart, Sustainable, and Inclusive Growth in the European Union. In M. Vemić (Ed.), *Optimal Management Strategies in Small and Medium Enterprises* (pp. 268–292). Hershey, PA: IGI Global. doi:10.4018/978-1-5225-1949-2.ch013

Taghizadeh, S. K., Rahman, S. A., & Ramayah, T. (2017). Innovation-Driven Planned Behaviour Towards Achieving the Wellbeing of the Malaysian SMEs. In N. Ahmad, T. Ramayah, H. Halim, & S. Rahman (Eds.), *Handbook of Research on Small and Medium Enterprises in Developing Countries* (pp. 280–296). Hershey, PA: IGI Global. doi:10.4018/978-1-5225-2165-5.ch013

Thavinpipatkul, C., Ratana-Ubol, A., & Charungkaittikul, S. (2016). Transformative Learning Factors to Enhance Integral Healthy Organizations. *International Journal of Adult Vocational Education and Technology*, *7*(1), 65–83. doi:10.4018/IJAVET.2016010105

Theodoridis, A., Ragkos, A., Angelidis, P., Batzios, C., & Samathrakis, V. (2017). Typologies as Management Tools: Understanding the Environmental Attitudes and Economic Prospects of Mussel Farmers in Greece. In J. Stanković, P. Delias, S. Marinković, & S. Rochhia (Eds.), *Tools and Techniques for Economic Decision Analysis* (pp. 170–180). Hershey, PA: IGI Global. doi:10.4018/978-1-5225-0959-2.ch008

Thongpoon, S., Ahmad, N. H., & Mahmud, I. (2017). Sustainable Approach towards Thai SMEs: The Effects of Country Philosophy of Sufficiency Economy and Government Support. In N. Ahmad, T. Ramayah, H. Halim, & S. Rahman (Eds.), *Handbook of Research on Small and Medium Enterprises in Developing Countries* (pp. 1–22). Hershey, PA: IGI Global. doi:10.4018/978-1-5225-2165-5.ch001

Timilsina, B. (2017). Overcoming the Barriers of Strategic Planning, Implementation, and Monitoring in Turbulent Business Environment: A Qualitative Study on Finnish SMEs. In M. Vemić (Ed.), *Optimal Management Strategies in Small and Medium Enterprises* (pp. 226–248). Hershey, PA: IGI Global. doi:10.4018/978-1-5225-1949-2.ch011

Tongiani, M. G., & Zhao, S. (2017). Traditional Italian Food Products on the Chinese Market: Best Practices for Italian Small and Medium Enterprises. In M. Vemić (Ed.), *Optimal Management Strategies in Small and Medium Enterprises* (pp. 22–43). Hershey, PA: IGI Global. doi:10.4018/978-1-5225-1949-2.ch002

Torlak, N. G. (2016). Improving the Role of Organisational Culture in Change Management through a Systems Approach. In A. Goksoy (Ed.), *Organizational Change Management Strategies in Modern Business* (pp. 230–271). Hershey, PA: IGI Global. doi:10.4018/978-1-4666-9533-7.ch012

Tran, B. (2016). Gendered Social-Networking Organizations: A View of the Sexed Mentorship Relationships. *International Journal of Organizational and Collective Intelligence*, *6*(2), 26–49. doi:10.4018/IJOCI.2016040103

Uyen, N. T., & Zainal, S. R. (2017). Non-State SMEs in Vietnam: Understanding Strategic Situation and Implications for Organizational Performance. In N. Ahmad, T. Ramayah, H. Halim, & S. Rahman (Eds.), *Handbook of Research on Small and Medium Enterprises in Developing Countries* (pp. 297–321). Hershey, PA: IGI Global. doi:10.4018/978-1-5225-2165-5.ch014

Vargas-Hernández, J. G. (2016). Critical Analysis of the Influence of Transnational Capitalism on Institutions and Organizations. In R. Das (Ed.), *Handbook of Research on Global Indicators of Economic and Political Convergence* (pp. 237–253). Hershey, PA: IGI Global. doi:10.4018/978-1-5225-0215-9.ch011

Vargas-Hernández, J. G. (2016). Strategic-Spatial Analysis of the Implementation of Business Opening Politics of Mexico. In R. Das (Ed.), *Handbook of Research on Global Indicators of Economic and Political Convergence* (pp. 274–313). Hershey, PA: IGI Global. doi:10.4018/978-1-5225-0215-9.ch013

Vargas-Hernández, J. G., Muratalla-Bautista, G., & Cruz, I. D. (2017). Point of View of Economical Organization at Manzanillo's Harbor. In R. Das (Ed.), *Social, Health, and Environmental Infrastructures for Economic Growth* (pp. 307–324). Hershey, PA: IGI Global. doi:10.4018/978-1-5225-2364-2.ch015

Vemić, M. B. (2017). A Further Look at Working Capital Optimization in Medium-Sized Firms: Concepts and Evidence. In M. Vemić (Ed.), *Optimal Management Strategies in Small and Medium Enterprises* (pp. 144–177). Hershey, PA: IGI Global. doi:10.4018/978-1-5225-1949-2.ch008

Vemić, M. B. (2017). Financial Innovation in Medium-Sized Enterprises Optimizes Their Gravitation Towards Capital Markets: Financial Future in Perspective. In M. Vemić (Ed.), *Optimal Management Strategies in Small and Medium Enterprises* (pp. 198–224). Hershey, PA: IGI Global. doi:10.4018/978-1-5225-1949-2.ch010

Villegas, M., & McGivern, M. H. (2015). Codes of Ethics, Ethical Behavior, and Organizational Culture from the Managerial Approach: A Case Study in the Colombian Banking Industry. *International Journal of Strategic Information Technology and Applications*, *6*(1), 42–56. doi:10.4018/IJSITA.2015010104

Virkar, S. (2016). Can Codes of Ethical Conduct Work?: Evaluating the Effectiveness of Privatised Corporate Governance in a World of Political and Economic Convergence. In R. Das (Ed.), *Handbook of Research on Global Indicators of Economic and Political Convergence* (pp. 206–236). Hershey, PA: IGI Global. doi:10.4018/978-1-5225-0215-9.ch010

Windsor, D. (2017). Defining Corporate Social Responsibility for Developing and Developed Countries: Comparing Proposed Approaches. In D. Jamali (Ed.), *Comparative Perspectives on Global Corporate Social Responsibility* (pp. 1–27). Hershey, PA: IGI Global. doi:10.4018/978-1-5225-0720-8.ch001

Wolf, R., & Thiel, M. (2017). CSR in China: The Road to New Sustainable Business Models. In D. Jamali (Ed.), *Comparative Perspectives on Global Corporate Social Responsibility* (pp. 258–275). Hershey, PA: IGI Global. doi:10.4018/978-1-5225-0720-8.ch013

Wronka-Pośpiech, M. (2017). Applying Business Solutions to Social Problems: Social Co-Operative and Its Business Model – Evidence from Poland. In M. Lewandowski & B. Kożuch (Eds.), *Public Sector Entrepreneurship and the Integration of Innovative Business Models* (pp. 139–164). Hershey, PA: IGI Global. doi:10.4018/978-1-5225-2215-7.ch006

Wynn, M. G., & Brinkmann, D. (2019). Developing Strategic Capability through Business Intelligence Applications: A case study from the German Healthcare Insurance Industry. In W. Yeoh & S. Miah (Eds.), *Applying Business Intelligence Initiatives in Healthcare and Organisational Settings*. Hershey, PA: IGI Global; doi:10.4018/978-1-5225-5718-0

Yap, N. T., & Ground, K. E. (2017). Socially Responsible Mining Corporations: Before (or in Addition to) Doing Good, Do No Harm. In D. Jamali (Ed.), *Comparative Perspectives on Global Corporate Social Responsibility* (pp. 185–207). Hershey, PA: IGI Global. doi:10.4018/978-1-5225-0720-8.ch009

Yordanova, S. D. (2017). Optimizing Virtual Communities in Tourism to Facilitate Development of Small and Medium-Sized Enterprises. In M. Vemić (Ed.), *Optimal Management Strategies in Small and Medium Enterprises* (pp. 60–78). Hershey, PA: IGI Global. doi:10.4018/978-1-5225-1949-2.ch004

Yusoff, A., Ahmad, N. H., & Halim, H. A. (2017). Agropreneurship among Gen Y in Malaysia: The Role of Academic Institutions. In N. Ahmad, T. Ramayah, H. Halim, & S. Rahman (Eds.), *Handbook of Research on Small and Medium Enterprises in Developing Countries* (pp. 23–47). Hershey, PA: IGI Global. doi:10.4018/978-1-5225-2165-5.ch002

Zel, U. (2016). Leadership in Change Management. In A. Goksoy (Ed.), *Organizational Change Management Strategies in Modern Business* (pp. 272–288). Hershey, PA: IGI Global. doi:10.4018/978-1-4666-9533-7.ch013

About the Author

Martin Wynn is Reader in Business Information Systems in the School of Business and Technology at the University of Gloucestershire, UK. He has a track record of research and publications dating back to the late 1970s as well as almost 20 years in industry as an IT professional and senior manager at Glaxo Pharmaceuticals Ltd and HP Bulmer Drinks Ltd, where he was IT Controller for 10 years. Since his return to academia in 2002, he has focussed on developing knowledge transfer activities for the University and played a major role in the leadership and management of this "third stream" activity. He has won, designed and secured funding for over 40 Knowledge Transfer Partnership projects in a range of disciplines – computing, marketing, new product development and environmental science. He was himself the academic supervisor or lead on 20 of these projects and it is these that are featured in this book. In addition to technology transfer, his current research interests include information systems strategy development, risk management and sustainability. He is also a successful supervisor of doctoral students, a qualified project manager and certified management consultant.

Index

V

W

www.ingramcontent.com/pod-product-compliance
Ingram Content Group UK Ltd.
Pitfield, Milton Keynes, MK11 3LW, UK
UKHW051028041026
15470UKWH00015B/1648

9 781522 574088